"While Kimberly Roberson's book is an insightful and well-written account oɪ ʟɪɪᴄ global consequences we all face from the ongoing disaster at Fukushima, it is also much more than that. Her story is an inspiring example of the power of taking action and getting involved. Told from a mother's perspective, Kimberly's book is about uncovering the truth while swimming upstream against governmental lies that have become almost cliché, and an incognizant public that's either too disinclined or afraid to seek the truth. Her book is a profound reminder that knowledge is power, that ordinary people are the ones who often change the world, that it is better to light a candle than to curse the darkness, and that we all have the power to provoke solutions to our greatest problems by taking action and inspiring others to do the same."

   – **Bill Richardson**, Deputy Executive Director, Greenpeace, USA

"Governments underestimate the effect of official silence on anxious mothers! Kimberly is not alone, even though she is more knowledgeable than many others, in becoming increasingly concerned about the Fukushima-Daiichi nuclear disaster in Japan and its effect on California milk, water, spinach, mushrooms and other foods. Platitudes about 'permissible doses' cannot take away parental responsibility and non-monitoring of fallout by the government does not make the problem go away! Thank God for mothers like Kimberly! I encourage everyone to read this small book with reliable information written in ordinary lay language. You will learn how passive most Americans were during this major disaster. You will learn how negligent were the government agencies we depend on for meaningful and timely warnings and advice for self-protection and caring for one's family!"

   – **Rosalie Bertell**, PhD, GNSH founder of the International Institute of Concern for Public Health (IICPH) and former consultant to the Nuclear Regulatory Commission, the Environmental Protection Agency, Health Canada

"We all owe Kimberly Roberson a debt of gratitude for bringing to light the health and safety issues related to the dangerous radiation releases from Fukushima. She underscored the need for food monitoring of radiation and transparency of findings in the U.S. food supply, even if U.S. elected officials and the media choose to ignore the ongoing problem. This is a 'must read' for anyone concerned about the impact of radiation on future generations."

– **Wenonah Hauter**, Executive Director, Food and Water Watch

"Silence Deafening is a must read! Ms. Roberson has identified the radioactive elephant in the global room that is ruining our health and REALLY threatens our collective survival. This is not a story by the girl who cried wolf. I appreciate her facts, feelings, food sensibility and leadership on how to advocate for an end to nuclear power. Message: don't be scared, be prepared. From the ground up and from the top down, radioactivity is a growing concern that must be measured and can be curtailed once we get out of shock and address a root cause of our damaged and dying civilization, toxic radiation."

– **Ed Bauman**, PhD, Founder and President of Bauman College: Holistic Nutrition and Culinary Arts

"Kimberly Roberson's wonderful new book is engaging, exciting and eloquent. No one can speak better to the radioactive crisis threatening our Mother Earth than a committed activist mother. This is an important contribution to the Solartopian dialogue, a graceful addition to the litany of going green …"

– **Harvey Wasserman**, Author of *Harvey Wasserman's History of the United States,* and *Solartopia!: Our Green-Powered Earth AD 2030*

"I highly recommend you read (and heed) Ms. Roberson's account, 'Silence Deafening, Fukushima Fallout … A Mother's Response.' This writing combines narrative, investigative journalism and whistle-blowing in informative prose. I started out doubtful, worn (what, another crisis I have to know about?) but Kimberly's self-reflective style of identifying herself as a 'stay-at-home mom,' her anecdotal asides, plus the information she offers along the way dusted off the doubts and whittled away the weariness. I'm glad I now know about this ignored disaster, and will attempt to be a force for knowing and no longer a force for ignoring. Join me, and Ms. Roberson, to disseminate this information."

– **Dr. Susan Parenti**, Co-Author with Dr. Patch Adams *The Politics of Care* and Director of School for Designing a Society

"For new mothers who are too young to have been educated in the horrors of the nuclear above-ground testing age as I was in the 1960s, this is an essential read. In her 20s, Kim's passion for a healthy planet brought her together with us to work against the irresponsible handling of radioactive waste as we battled against putting it into holes in the ground that would have contaminated the water supply for millions. Now, as a mother of a young son, her passion again compels her to raise awareness of the enormous and unspoken results of Fukushima fallout. Hopefully mothers and grandmothers everywhere will join with Kim and the mothers in Japan to stop this nuclear insanity for everyone and the children of the future."

> – **Marybeth Brangan**, Filmmaker and Co-Director, Ecological
> Options Network

`"If more mothers like Kimberly Roberson act on their concerns, we can go a long way to reducing future threats of nuclear power contamination of our food supplies. How dare this industry disperse so much known cancer and other disease-causing radioactivity locally and globally, endangering this and future generations, all the while denying the hazard? It's pretty hard to avoid the invisible but real poisons still being emitted from Fukushima (and every nuclear power reactor) when we cannot see, smell, taste or feel them and when the government that licenses this industry refuses to provide or require meaningful monitoring or reporting. Thanks Kim for learning all you can and for yelling so loudly and clearly."

> – **Diane D'Arrigo**, Radioactive Waste Project Director, Nuclear
> Information and Resource Service

"The burden of knowing too much is all to evident in 'Silence Deafening, Fukushima Fallout … A Mother's Response' by Kimberly Roberson. A person can do two things with this burden: they can choose to ignore what they know and go about life as it is now, or they can ACT. Roberson has chosen to act because not doing so was unconscionable to her. This is clear in her well-researched and passionately written book which seeks not only to share important knowledge of radiation contamination in the wake of the Fukushima nuclear catastrophe in Japan, but also to convince people knowledge is power, they do have a voice and can take action to protect themselves and the ones they love. The ultimate answer: shut the nuclear reactors down and usher in a cleaner, friendlier energy future with wind and solar power at its core."

> – **Cindy Folkers**, Radiation Health Hazards Specialist,
> Beyond Nuclear

SILENCE DEAFENING

Fukushima Fallout ... A Mother's Response

Kimberly Roberson

"To my anti-nuclear heroes ... past, present and future."

Published by

VisionTalk™

Playa Del Rey, CA

Copyright 2012 Kimberly Roberson

First Edition 2012

ISBN 978-0-9853412-0

Table of Contents

FOREWORD

I awoke the morning of March 11, 2011, knowing that something was terribly wrong. From within, all I could feel was overwhelming and unbearable grief. It was as if the voice from inside my head had expanded exponentially into a sea of voices crying out in pain and sorrow.

Upon checking my phone I saw a message -- "Earthquake in Japan, Tsunami ..." This was all I could bear to read. The anguished voices from within grew louder and my upset and grief continued to grow worse ... I did not even bother to fight back the tears. 'Was I going crazy?' 'How could I be experiencing so intensely these feelings in connection with the events from the other side of the world?' In reflecting later, I feel I somehow tuned into the "what" that was happening; not just with Japan and the Japanese people but also with our Earth through this process. Mother Earth was screaming in agony and somehow I had tuned in directly to her frequency.

As the dust settled, news came in of the Fukushima Daiichi disaster, and with it came the awareness of the radiation that was being poured into our atmosphere and oceans. I immediately began looking for what to do to protect myself from this contamination. Amazingly I found almost nothing! I knew that "Someone, some-where must know what to do!" I realized in this moment that I was committing to this cause ... whatever it looked like.

Flash forward six months later to September 2011. A friend suggested that I might like to speak with a mom who wants to create a book about Fukushima ... And in walks Kimberly Roberson into my life ...

Kimberly and I came together to publish this book, a "field guide, " which is Kim-berly's response to the events at Fukushima Daiichi that began on March 11, 2011.

This book is a model of an awakened mind, body and spirit and in action. Kimberly takes all that she knows from her experience and expertise, and applies this with all of her heart and ability in service to our education, empowerment, and in that we all have the right to "know" what to do to support ourselves and our families in response to this disaster. Kimberly has done the research, walked the walk through her own life experiences that have brought her fully into the here and now. She takes us through her very personal journey of seeking to provide clean food and water for her family, as well as insisting that our government and its agencies not only begin monitoring our food and water for radiation levels to ensure we are well informed of how the radiation of Fukushima will effect our food and water sources in times to come, but in also seeking a cessation of nuclear power full stop.

With all of this said, I am grateful for the opportunity to publish and distribute this book and to support Kimberly and her mission to educate and share the truth about what is happening, and what we can do to change this current and future reality.

If I am infinitely sensitive enough to feel, from halfway around the world, the devastation, anguish, travesty and tragedy of those who were involved with the earthquake, tsunami, and following nuclear meltdown that led to this level of suffering, I am also infinitely powerful enough to be a part of changing this reality in some way. We ALL are.

This book is just that. An opportunity for change, and for us to each to wake up to the truth about what is happening in our world and to commit to doing our part in support of the solution, much as Kimberly has done in sharing her story in this book.

This book is also an invitation. You have already answered the call. Now as you ready to embark on this journey and as you read this book, I encourage you to be thinking about what it is you have to contribute.

It takes each and every one of us to do our part to find solutions for these challenges that face us all. We each individually hold our piece of the global peace puzzle. Kimberly's piece is her voice and her unique perspective, and her commitment and ability to impact change through sharing her story and experience.

The time is now, there is an urgency, we ARE determining our destiny in this and every moment! –

– Melinda Woolf, Vision Talk

We ourselves feel that what we are doing is just a drop in the ocean. But the ocean would be less because of that missing drop.

- Mother Teresa of Calcutta

1

Before Fukushima and After

One nuclear accident can ruin your whole day.
– Unknown

Now more than ever it's becoming clear to me that humor and irony help to emotionally deal with chronic, constant awareness of chronic, constant radiation. I've come to think of March 11 and 12, 2011, as *BF* and *AF, Before* Fukushima and *After*. My son's third birthday was at the end of February, *BF*. When pouring over pictures from his jungle-themed party I'm struck by a loss of innocence, and amazed at how we never know how good we have it — until we don't.

Fukushima Daiichi is widely considered to be the worst nuclear disaster in world history and there is no end in sight as of this writing. The International Nuclear Event Scale (INES) lists Fukushima at a Level 7, the highest possible on the rating scale, along with the Chernobyl nuclear disaster of 1986. There is no Level 8, however with radiation releases continuing from Fukushima and no permanent containment in sight, one has to wonder if that might change.

The silence after the earthquake, nuclear meltdowns and tsunamis *AF*, was truly deafening and unlike anything I had experienced before. Surreal, "Twilight Zone" comparisons were hard to avoid. Knowing what I knew, and then seeing those facts be so thoroughly disregarded by the media and elected officials had begun to take on a sort of nightmare quality. What most certainly intensified the crisis at Fukushima Daiichi for me was the fact that a typical suburban mom doesn't usually list the Greenpeace Radioactive Waste campaign on her resume — who then went on to study and work in the field of holistic nutrition before starting a family. I have always considered it to be a pretty unusual leap from radioactive waste to nutrition and then to motherhood, but one day those three worlds would merge closely, on March 11, 2011, to be exact. The one thing such experience brings that is startlingly clear to me: radioactive fallout from nuclear power and food do not mix, and children are especially at risk.

By the latter part of 2011, a much more complete truth of the catastrophic events of March 11 were slowly becoming known to the world via experts, concerned citizens and the Internet. While the world was riveted by images of the earthquake and ensuing tsunami, a third and more horrific event was unfolding which very few people understood at the time – one which many still don't fully understand to this day. It may take decades for the true magnitude of Fukushima Daiichi to be fully comprehended, just as the effects of the Chernobyl nuclear meltdown of 1986 are still being realized. This is the story of my attempt to learn the truth, and then to do something about it in my own small way.

Things don't always turn out as we plan. Sometimes life is more like a trajectory honing in on the focus of why we are meant to "be here." After leaving Greenpeace I was invited to live and work on an idyllic organic farm in Northern California where I continued to work with safe energy activists who inspired me beyond measure. I embraced the natural world and its many delights, including learning everything I could about eating in a healthy way. This eventually led to studying holistic nutrition in San Francisco and then working in that field for nearly a decade. Everything changed when my husband and I decided to start a family. From this point on it was all about our child. Parenthood changes everything, another cliché, and I would joke to friends that Planet Me became Planet Son. When Fukushima Daiichi struck it was a logical step for me to need to do something to help break the silence to protect him and his little friends.

I began the motherhood journey later in life, which alone has its pros and cons: you know more, but then you *know more*. So a somewhat unique skill set came in handy while writing an on-line petition calling for monitoring and transparency of an increasingly radiated food and water supply in the U.S. Even in the era of petition fatigue, this became one way that a stay-at-home mom could spread the word about the need to monitor food and water during and after the Fukushima Daiichi nuclear crisis — whenever "after" actually will be, we can only guess. Collecting signatures on the Fukushima Fallout Food Safety Petition turned out to be a frustrating process because it seems most people prefer avoiding such thoughts and believe that their government will do the work for them. They are wrong.

Those who cannot remember the past are condemned to repeat it.

– G. Santayana

Most mothers don't give nuclear power a second thought, but I had a history, you see, going back to my early activist days. I started in 20s, when most people believe they can change the world if inclined to such things. When a job opened up on the Greenpeace East West Campaign I jumped at the chance to work in DC as an administrative liaison to the director who was opening the first office in Moscow. This was in 1989, and I still remember opening the letter from the farmer near the Chernobyl nuclear disaster who had mailed us pictures of grossly deformed farm animals. Those images would later appear in magazines like Time and Newsweek and helped to open the world's eyes to the largest nuclear disaster to date. Eventually the job transitioned and I decided to go west. As fate would have it, I was working in LA when news broke about plans for the Ward Valley radioactive dump site. This was basically a proposal by the nuclear industry to dig unlined dirt trenches directly above a major underground aquifer leading directly to the Colorado River in Southern California — the agricultural and drinking water supply source for much of the western region of the U.S. Activists working to block the proposed dump learned quickly that there is in fact no such thing as "safe" low-level radiation. We would come to learn that there are vast misconceptions about what actually constitutes "low-level" radioactive waste — the term itself is seriously misleading.

Like other moms in my circle, I know more than a few children under the age of three with cancer and two of them recently died. This is alarming because I don't remember even one child who had cancer when I was growing up. Now, we know of one who was a precocious little boy whose parents were environmentalists with healthy lifestyles. They did everything they could to provide the best for him, only to see fragile his life slip away due to cancer. Another young child we know is relapsing with leukemia and is enduring another long round of painful chemotherapy while awaiting a bone marrow transplant. And it seems we all have family and friends with stories of their own courageous battles. Cancer seems to have become a sort of bizarre birthright for so many people, young and old, that it's reached epidemic proportions in our industrialized world. And while this may seem anecdotal, there has definitely been a surge in cancers of all types in the years since I was a child.

When I decided to pursue a career in holistic nutrition, the decision provided a chance to become healthier myself, an added bonus. My colleagues and I enjoyed daily connections to the earth through food that such awareness provides. It's one of the few ways that we stay connected to nature on a continuous basis if we are mindful. A healthy environment means a healthy diet, and vice versa, they are interdependent. Forget about "saving the planet." Mother Earth will exist for millions of years even if it means shaking humans off her back. It's more about saving ourselves when the bottom line is truly understood.

2

You Can Never Be Too Prepared

Try having a conversation about all of this around the swings at the neighborhood park, where one can now realistically wonder now how much radioactivity from Fukushima Daiichi collected from the rainy season in the sandbox. After the initial Fukushima explosions, parents in Japan took matters in their own hands by renting backhoes to shave radioactive topsoil from the playgrounds there. It's tricky telling neighbors downstream here why your child won't be swimming in the community pool when its 100 degrees outside and their children shouldn't either; that radioactive strontium, cesium and hundreds of other hazardous radioactive isotopes could have possibly collected in open swimming pools during an especially long rainy season which lasted for over three months *AF*. I've become accustomed to the shocked looks, then the slow realization that most people don't want to hear and certainly don't want to believe radioactive fallout could be happening to them.

BF, worry during motherhood was something that happened pretty much most of the time anyway. Just mindfulness, I would tell my husband, albeit a stressed out version. This is what happens when your previous profession taught you that toxins from smoke stacks are routinely measured in tons, that the average fetus at nine months has over 200 chemicals in their tiny body in utero, that roughly 80,000 chemicals have been created for use in our daily lives and that then go unmonitored regardless if they are carcinogenic — well, it makes mommy fear grow exponentially, and not your usual garden variety of mommy fear at that. Environmental activism has offered numerous opportunities for reflection on "total body burden" of toxins.

When I was pregnant and considering breastfeeding I couldn't help but wonder if it would be safe for my tiny son. A burden indeed! A dear friend who understood my concerns sent Sandra Steingraber's book *Having Faith, an Ecologist's Journey to Motherhood*. In it she ponders the health of her unborn daughter Faith and whether the toxins in her own body could harm her daughter. Ms. Steingraber is an accomplished biologist who presented to the United Nations as part of international treaties meetings and carried along a vial of Faith's sustenance to present during her speech on the safety of human breast milk. In 2007, when this book was written, she declared in her estimation that the scientific benefits outweighed the risks, but just barely. It seemed like a gamble, but if it were good enough for her child, hopefully it would still be safe to nurse mine. *Having Faith* had quite an effect on my outlook, so much so that I embarked on a personal campaign to avoid chemicals in body lotion, shampoo, cosmetic and nail polish, and choosing only locally organic and fresh produce and dairy. We ordered a mattress wrapper on-line from New Zealand to help ensure that our baby did not inhale possible fumes from his bed that could potentially lead to SIDS. We used corded phones in our home, and our cellphones were turned off. Bases covered? Had I created as safe nest? One would hope. We didn't know the challenges that were to come...

3

"Criticality" — The New Awareness

But that was *BF*, before Fukushima. Three years later, in April 2011, *AF*, Tokyo Electric and Power Co. (TEPCO) announced that it would take well into 2012 or beyond to stop the radioactive fallout of the worst nuclear disaster in world history. Now we know that that number was inaccurate just like many other statements issued by the utility company concerning the disaster (much more on that later).

From the beginning of the disaster officials for the nuclear industry were assuring the public that Fukushima was not Chernobyl, as though this was something in which to take comfort. Chernobyl, prior to Fukushima, was the largest nuclear power disaster in world history that began on April 23, 1986. Some industry sources claimed that 60 people died due to injuries from Chernobyl, and the consequences at Fukushima would also be negligible. However the New York Academies of Science report dated April 26, 2010, stated that of the 600,000 Soviet soldiers and citizens that were recruited to clean up the Chernobyl disaster, nearly 40,000 died as a result of their injuries due to radiation, and 200,000 troops were left incapacitated. All told, an estimated 1,000,000 people around the world died a result of Chernobyl.

Perhaps the biggest tragedy is the impact on the children who were affected while in their mothers' wombs. Hundreds of orphanages in the Ukraine were filled with children who suffered mental and physical deformities due to the radiation from Chernobyl. Another difference between Chernobyl and Fukushima is that Chernobyl released radiation directly over land before spreading widely over Europe and out to the Atlantic, where it then circled the planet many times. Fukushima did somewhat the opposite, releasing radiation directly over the Pacific Ocean after first spreading south after two large explosions, several days apart. Atmospheric maps showed that the plume rode the powerful jet stream and was detected reaching U.S. shores seven days later on March 18.

At Chernobyl, there was one reactor affected, at Fukushima there are four, and workers are still struggling to contain radiation there as of this writing nearly one year after the disaster began March 11. The incredibly brave Soviet workers at Chernobyl — many of them soldiers and hard rock miners who gave their lives while working to contain much of the widespread radioactive fallout — dug under the reactors and then built a barrier to avert a "China syndrome" (where the core continues to "meltdown" into the earth).

The Chernobyl disaster was in 1986 and by 2011 experts realized that the aging sarcophagus (the above-ground structure built to contain the fallout) was aging badly, degrading to the point that it threatens to this day to release spent fuel into the atmosphere once again. Nuclear power seems like a bad idea all around because the very function of nuclear reactors requires a daily, continuous release of radiation into the biosphere. I learned that the very action of splitting atoms creates a sort of ongoing criticality that is controlled to a degree within the reactor walls. If something goes wrong (and there are literally a hundred or more relatively minor accidents at nuclear power plants each year) well, it can go wrong badly. One of the biggest tragedies of Fukushima is that tens of millions of gallons or more of radioactive plutonium and strontium fuelled water were dumped into the ocean during TEPCO's attempt to use fresh ocean water to cool the damaged reactors. The Canadian Food Inspection Agency has stated that approximately 60 per cent of fish have shown to have detectable levels of radionuclides, however their government still has not acted to protect citizens there. Meanwhile in Japan, 60 to 80 percent of fish caught in Japanese waters each month are found to contain radioactive cesium, as tested by Japan's Fisheries Agency. So many lessons ignored by the powers that be. Why would we want to attempt to continue to harness a form of energy over which we can't control?

Much of truth of Fukushima was apparent by the end of 2011. However merely 10 days after the disaster struck the well-respected French NGO, Commission for Independent Research and Information on Radioactivity (CRIIRAD), issued a statement that radiation from Fukushima was "no longer negligible" and advised French pregnant and nursing mothers and young children to avoid large leafy vegetables and soft cheeses. China and India also cautioned their citizens to carefully handle or avoid large leafy greens. Chris Busby, PhD, Scientific Secretary of the European Committee on Radiation Risk, published a *"Don't Panic"* guide in early April, 2011, saying that the danger was insignificant, but later changed his position stating *"since then I have re-thought this advice as the thing is still fissioning. This will mean that radioactive strontium uranium and particulates will be building up in the USA and Europe. I will assess this later but for now I think it prudent to stop drinking milk."*

Here on the West coast of the U.S. and closer downwind of Japan, there was no word from our government whatsoever other than the radiation was all within "safe range" and "to be expected." No mention of alpha, beta or gamma radiation for that matter. Just rest easy. And all along while reading up on this stuff I kept thinking, *"TEPCO Fiddles While the World Burns."* Not everyone was fooled, however. Dr. Rosalie Bertell, PhD and GNSH (Grey Nun of the Sacred Heart) is an accomplished scientist who has dedicated her life to unraveling the mysteries of ionizing radiation and its impact on human health. Dr. Bertell has written about the topic of double-strand DNA damage as a consequence of radioactive fallout. In her *No Immediate Danger, Prognosis for a Radioactive Earth* she warns not only of the damage to the person coming in contact with radioactive fallout in their food and water, but also to their children, grandchildren, and great grandchildren suffering mutations in their DNA as well. She is just one of many scientists who understand the true repercussions of the nuclear industry.

4

Alpha, Beta, gamma — A Primer

Well into my campaign I received an unexpected call from a kind and knowledge-able woman named Cindy Folkers from Beyond Nuclear, a Washington, DC-based watchdog group. Cindy had seen my petition on-line and was concerned that the language didn't differentiate one form of ionizing radiation from another, and that she feared that the Japanese were only testing for some but not all types. And it was true indeed I needed a refresher course, although it is not rocket science and all of us should understand the basics of radiation.

It helps to have an understanding of what radiation is, as it can be stored cumulative in the body. Alpha, beta and gamma are all types of radiation that emit energy differently, all of which are hazardous when inhaled and/or ingested.

Alpha are electrically charged and the least penetrating of the three because they travel between four to seven inches in air and can be stopped by a sheet of paper or the outermost layer of skin that covers the body. However, if an alpha particle-emitting radioactive material is inhaled or ingested, it can be an extremely damaging source of radiation because it can still be concentrated internally. Once lodged in tissues internally, it continues to give off radioactivity to the surrounding cells.

Beta particles travel faster and penetrate deeper than alpha, at several millimeters through tissue. Like alpha particles, the greatest threat is if beta particle-emitting radioactive material is inhaled or ingested.

Gamma rays are similar to X-rays, and are a form of electromagnetic radiation. They are the most hazardous type of external radiation, as they can travel long distances and penetrate all types of materials. Gamma rays penetrate more deeply through the body than alpha or beta, and all tissue and organs can be damaged.

When considering that all of these are produced during nuclear power production and that some can effect future generations, we have to wonder how this form of energy became so widespread in the first place.

The following, a more in-depth account of how radiation works, is from the Beyond Nuclear fact sheet that is available at www.beyondnuclear.org:

RADIATION BASICS

What is radiation? Radiation is energy that travels in waves or particles. Each type of radiation has different properties. Non-ionizing radiation can shake or move molecules. Ionizing radiation, the kind expelled from nuclear reactors, can actually break molecular bonds in our cells, causing unpredictable chemical reactions. Humans cannot see, feel, taste, smell or hear ionizing radiation. Unavoidable exposure to ionizing radiation comes from cosmic rays and some natural material like rocks or soil. Human exposure to natural radiation is responsible for a certain number of mutations and cancers. Any additional exposure above natural background radiation can result in otherwise preventable disease since there is no safe dose.

Radiation damage and protection levels are based on "Reference Man," a healthy, white male in the prime of life, and mostly ignore the more vulnerable fetus, growing infant and child, the aged, those in poor health, and women who are, according to the National Academy of Sciences, 37 to 50 per cent more vulnerable than adult men to the harmful effects of ionizing radiation. These levels, therefore, do not take into account the far greater vulnerability of women and children, especially pregnant women and unborn children. Further, a panel from the U.S. National Academy of Sciences (NAS) charged to investigate the dangers of low-energy, low-dose ionizing radiation has concluded, "that it is unlikely that a threshold exists for the induction of cancers ..." (BIER VII, 2005).

Therefore, saying that there can be a "safe" level of radiation exposure is simply wrong. There is no guarantee that even the smallest doses of radiation will not cause harm.

How Radiation Harms

Ionizing radiation travels through our living tissue with much more energy than either natural chemical, or biological functions. This extra energy tears mercilessly at the very fabric of what makes us recognizably human – our genetic material.

11

Elderly and people with immune disorders are more susceptible to ionizing radiation. Women are more susceptible to this damage then men and children more susceptible than adults. Children and the unborn are especially susceptible because of their rapid and abundant cell division during growth. Female children are the most susceptible.

Cancers linked to ionizing radiation exposure include most blood cancers (leukemia, lymphoma) lung cancer and many solid tumors of various organs. Heart ailments are also associated with radiation exposure.

Additionally, evidence exists that radiation is permanently and unpredictably mutating the gene pool and contributing to its gradual weakening. The New Scientist quotes a report that calls genetic or chromosomal instabilities caused by radiation exposure a "plausible mechanism" for explaining illnesses other than cancer, including "developmental deficiencies in the fetus, hereditary disease, accelerated aging and such non-specific effects as loss of immune competence."

5

ATOMS FOR "PEACE"

Perhaps most people haven't heard about the "Atoms for Peace" program of the 1950s where, here in the U.S., nuclear weapons technology was converted to an energy source which was promised to be "too cheap to meter." This was an effort to convert weapons grade plutonium previously used to bomb Japan during WWII into a friendly source of energy that promised to transform humankind, but we could never have guessed what that transformation would truly mean. I've always remembered a quote from my early activist days passed down from former President Dwight Eisenhower. He spoke to his concern about nuclear power, *"How far can you go without destroying from within ... what you are trying to defend from without?"* I always took those words to mean that he understood the dangers of the new technology and feared the long-term consequences. Now, nearly 70 years later we are exacting that "too cheap" price more than ever before, with radioactive fallout hitting home in more ways than the jet stream. The Fukushima Daiichi reactors, I learned, are the same GE Mark I model as many here in the United States. Although Diablo Canyon and San Onofre in Southern California are different in design, however, both are on the shore and directly on or near active earthquake faults just as Fukushima. And the threat to human health is clear because cancer rates spike in nuclear reactor communities, thyroid, breast and leukemia being among the highest percentages.

General Electric continued to build reactors around the world, going to places like New Mexico and Africa for the deadly raw material to fuel them. I recalled from days working on Ward Valley that uranium mining is an extremely dangerous process that produces mill tailings (a type of uranium waste byproduct). This is waste that remains deadly for millions of years.

Uranium mining has created not only nuclear power, but also a legacy of disease and death. The Navajo Indians have experienced some of the worst of this tragedy in the United States. Uranium mill tailings and piles were created due to widespread uranium mining beginning in 1940's up until the 80's. Millions of tons of uranium ore were mined from the Navajo Nation reservation spanning Arizona, New Mexico and Utah. Tragically, many of the Navajo people were unaware of the dangers and actually built their homes with chunks of uranium ore and mill tailings. Uranium poisoning has been directly linked to cancer due to radiation poisoning. This is a tragedy that few people talk about, however the Navajo people suffered with increased cancer rates in their communities and little hope of rebuilding on their own. The US government began to rebuild in 2009 by promising to pay up to 3 million dollars per year to destroy and rebuild affected structures on the reservation.

Nuclear proponents who claim that nuclear power is relatively clean compared to coal ("no smoke from the stacks") are completely disregarding the uranium mining component and the nuclear fuel chain. In fact plutonium, which is produced from the uranium to be used in the reactors, is the most dangerous element known to humankind. What could possibly be peaceful about that?

And all the while, Japan experienced strong aftershocks (one as recently as January 12, 2012, as of this writing) that continued to threaten the precarious situation at Fukushima Daiichi. Protecting the fuel pools from exploding is a major concern according to experts such as Arnie Gundersen, a former nuclear industry senior vice-president with 40 years of experience working on nuclear power. During his career he managed and coordinated projects at 70 nuclear reactors in the U.S., and he and his wife Maggie run Fairewinds and its website, which is one of the most comprehensive in the world regarding the realities and dangers of nuclear power. He has been one of the few experts speaking the truth in the mainstream media (his credentials are very hard to ignore). He was one of the first experts to sound the alarm about Fukushima, he asserted that the average Vancouver and Seattle resident was inhaling eight to 10 hot particles per day by April, 2011.

6

MY WAKE UP CALL

One way to open your eyes is to ask yourself, "What if I had never seen this before? What if I knew I would never see it again?"

– Rachel Carson

All of this bad news meant that I had to do something. Starting an on-line petition was akin to a lifeline for a stay-at-home mom such as myself. My husband knew about the anti-nuclear campaign work before we met, but how could he or most people for that matter possibly understand what people who worked on such things do? I would tell him that it wouldn't be so worrisome if it weren't for our son because young children are still developing and especially vulnerable. And what must the parents of Japan be feeling? A beautiful and sacred culture was now horribly stricken by yet another nuclear disaster. And what about Hawaii? Consider the magnificent paradise, with beauty beyond measure. Hawaii is still part of the United States last time I checked, however no word from our government about the significant radiation spikes and findings there which have led dairy farmers on the Big Island to spread boron on their land as a protective measure, just as was done in Europe after Chernobyl.

Grappling with my growing unease and depression was challenging, but working on the petition helped. Being mindful can be tough but so is parenthood in general. Raising awareness on this critical issue in a world already inundated with causes, crisis and concerns is not only timely but also crucial, especially for parents. Most of us think that it's fine and safe to give our kids milk to drink. But by mid-April, 2011, one could go on-line and find that iodine-131 was being detected in cow's milk in Spokane, Washington, and as far East as Philadelphia and even Vermont.

The idea for starting a petition began to take root a week or so after the earthquake and tsunamis. I was under no illusion that food and water could be protected by monitoring alone as called for in the petition, but was rather a place to start, with much work to undertake. In reality there is no way to completely turn back, the radioactive genie is out of the bottle and future generations will be responsible for care taking radioactive waste for tens of thousands of years. Embracing sustainability, reducing consumption, and phasing out nuclear power such as the Germans, Italians and Swiss are now doing as a result of Fukushima (and as China and several other countries seriously considering) is the only truly realistically effective course of action. The answer, unfortunately, is not immediate, and it is not as though we can switch from buying one product instead of another. But those of us who love our families possess the incentive and drive to make it happen. Ignorance may be bliss, but this doesn't produce results other than the status quo. And the clear message from Fukushima is that the status quo is no longer an option.

We need to remember the task of finding a way to safely store the millions of pounds of spent fuel created by nuclear power production. (I've always wondered about the term "spent fuel" because there is nothing spent about it: plutonium and uranium fissionable byproducts will remain hazardous for millions of years. What is "spent" about that?) Life as we know it must change. However such awareness is in short supply considering that President Obama has pledged billions of dollars for the proposed "nuclear renaissance." $18 billion has been approved as of this writing and all in the form of loans because Wall Street refuses to back a risky proposition such as nuclear power.

Leave it to the self-proclaimed "New Clear" industry to create a "Renaissance!" I had to laugh when I first heard that one. But thanks to the vigilance of dedicated environmentalists, humanitarians, activists, or all of the above and add hero to the title — thanks to them, for the past three decades, times are changing. At the root of it all, they are concerned people who set out to make the world a better place. And, in large part to their efforts, a 2011 Greenpeace report for the first time in history highlighted the fact that green technologies such as wind and solar power were beginning to gain in new installed capacity over coal and nuclear production worldwide. Germany is predicting to switch over to completely renewable energy by the midcentury. Clearly, there is a chance now to change the world for the better, our children's lives for the better. However if we choose instead to continue with a new generation of nuclear reactors, there will be NO turning back. President Obama's children drink milk. Michelle Obama's prized organic vegetable garden requires clean water to grow. It's hope against hope that logic will resonate and prevail, even in a massive uphill battle.

7

Our Silent Spring

Back in 1962 Rachel Carson, a researcher and biologist with considerable vision, wrote *Silent Spring,* a former New York Times bestseller and Discover Magazine's pick for one of the Twenty-five Greatest Books of All Time. In it she claimed that widespread pesticide use, which began post-WWII, was seriously threatening our environment, particularly birds, and that human health would also be gravely affected. *Silent Spring* highlighted ecology and in turn created the environmental movement that we know today. It also led to the complete ban of the pesticide DDT 10 years later. I wonder a lot about her now, and what would she be thinking today?

There is a creek behind our home and it was very apparent that the birds were not singing in the spring of 2011 as they had one year earlier. In 2010 it would often be necessary to take a broom outside to shoo them away from my son's bedroom window so they wouldn't wake him up during naptime. No such problem in 2011, "Why weren't the birds singing?" Dedicated activists and filmmakers Marybeth Brangan and Jim Heddle, co-directors of Ecological Options Network, or EON, had posted a YouTube interview with Dr. Dave Desanto of the Point Reyes Bird Institute almost immediately after the Fukushima disaster began. Dr. Desanto gave a chilling account of his research tracking the effects of Chernobyl on bird populations in the 1990s, research which continues to this day. I sat before the computer, riveted, as he recalled how he and his colleagues could find absolutely no reason for the drop off in bird populations then other than for Chernobyl. I can't tell you how much I missed the birds *AF*. In fact, I cried for them.

Apart from the occasional Internet bombshell, however, the deafening media silence around Fukushima raged on. It broke on rare occasion such as the day Food and Water Watch appeared on a search while frantically looking for any signs of information pertaining to food safety. They were in the process of a letter writing campaign asking for the monitoring of food and water in the U.S. due to Fukushima Daiichi. Having been out of touch with such issues for many years, their website was like a lifeline. So I wrote the petition using FWW as inspiration and worked several hours a day (early or late hours when my son was sleeping) using it to spread the word via Facebook, Twitter and mother's clubs (although a national "holistic" mom's club told me it was "too political" for them to post on their website).

The petition asks for Congress and President Obama for the following actions:

First: *to monitor all food and water imports from Japan, including the estimated annual five million gallons of bottled water, soft drinks and other non-alcoholic beverages containing water. Seafood shipments and other food products must also be monitored immediately.*

Next*, the Environmental Protection Agency must significantly expand the monitoring of air particulates, rainwater, drinking water and milk and to make the findings readily transparent and immediately available to the public.*

Last, *the United States Department of Agriculture and the Food and Drug Administration should receive funding for expanded food and water inspection both here and overseas and to communicate those findings immediately to the public. Congress must rethink our agricultural policies as well as international trade policies as they affect imports from other countries also trading with Japan.*

By Mother's Day there was "breaking news" — the widespread assertion by the Food and Drug Administration that the radioactivity at Fukushima Daiichi was comparable to flying cross-country on a commercial airliner. Again, no differentiation of gamma rays to alpha and beta particles was ever mentioned either by officials or the media. This only served to placate the masses, and was and still remains a gross and unethical disservice. If we did hear about any of the hundreds of radioactive isotopes, it was only iodine-131 (with a half life of eight days, meaning that its power is cut by half every eight days) but nothing of iodine-129 with a half-life of ***16 million years*** and hazardous for much longer than that. But all we heard of in April 2011 was iodine-131, half-life of eight days. No worries at all. By now I was e-mailing and speaking fairly regularly to experts due to the on-line petition (which was starting to gain some momentum). Many of them I knew from my activist days years before, and also meeting new experts who were also committed to keeping the public safe. It was hard for me to escape these truths where my contemporaries were still relatively unaware of the dangers. So when the 25th anniversary of Chernobyl came around, we were reminded of the fact that there is still grazing land in the United Kingdom that is still off limits to grazing due to cesium-137 contamination and its insidious 30-year half-life.

Meanwhile on TV, the Chernobyl anniversary was briefly mentioned, along with occasional reports cautioning against taking supplemental iodine too soon "before the plume hits." If and when the plume would hit no one knew at that time. Here in California iodine in oral form was hard to obtain because it was flying off shelves and on back order wherever listed on-line. Inconsistent reports led to a panic, and then more silence.

8

ATOMIC AGRICULTURE

Looking back now it's easy to remember the shock of those early days *AF* in the spring of 2011. Remember, *BF* I was content to be a stay-at-home mom to a very active child however *AF*, (post-311 as the Japanese were now referring to that terrible day) there was a profound awareness that the government was doing nothing to protect him or to educate me. Then one day, while flipping channels hoping for even a mere mention of the ongoing disaster, California Senator Barbara Boxer was in the midst of conducting a hearing televised by CSPAN regarding the safety of the two reactors in her state, San Onofre and Diablo Canyon, both built on the shoreline and either on or very near active earthquake faults. "You say it couldn't happen here," she admonished, "and that's exactly what you said about Japan!" Her presence via TV in my living room that day seemed like a true voice in the wilderness. Did she have the political will to follow through? Could she protect Californians and others around the U.S. from this ongoing radioactive fallout?

Being in the field of nutrition I remembered that California has been the nation's largest producer of food for nearly 50 years, but I was surprised to learn that California farmers produce more than 450 crops including just about any produce you can imagine. California is the United States' No. 1 producer of dairy products. In addition, California grows 80% of the nation's spinach and lettuce alone, not to mention almonds, walnuts and berries, all of it potentially affected by radioactive fallout from Fukushima Daiichi. If a nuclear disaster were to happen at either of the aging reactor complexes in California, well, then dire consequences would be a certainty. Enlisting Senator Boxer's help might prove to be a long shot, but a necessary one nonetheless. This was not only a state issue but also a federal one. I was hoping (rather naively perhaps) that President Obama would understand the urgency and act too.

I love the quote by German philosopher and acclaimed alchemist Goethe, who once famously said, *"Be bold and mighty forces will come to your aid."* We've all felt it: the mystical energy that carries you forward when fighting for something important. Events align and happen, as if somehow effortlessly when you're on the good path. But the petition had stalled at nearly 300 signatures in one month, and it was like pulling teeth to get even a friend to stop what they were doing to sign. And why should they? After all, it was like taking a transcontinental flight, according to reports. Where the hell were the mighty forces, the *good* kind? And perhaps even more maddening was that so many activist acquaintances seemed distracted, busy with their own pressing issues. One person whom I was sure would at least sign and maybe even circulate the petition wrote back, "I can't in good conscience sign this." Why? He wouldn't elaborate, but suspected it was because he was working on climate change issues and was perhaps one more smart person fooled into thinking that nuclear power was somehow better for the environment than coal. But just because *you can not see it or smell it, doesn't mean it isn't there.*

9

BEARING WITNESS

One of the founding principles of Greenpeace since their formation in the 1970s, of which has remained central to their core, is the act of *bearing witness*. Based on the Quaker philosophy, this means that once someone has witnessed an injustice, one cannot claim ignorance as a defense for inaction. One makes an ethical choice: to act or not to act. True to form for Greenpeace, they acted in the most ethical way possible by sending a crew over to Fukushima City in the days following the disaster to take radiation readings. Their findings shed considerable light on the situation and led officials in Japan to increase the International Nuclear Event Scale (INES) from a Level 5 to the highest possible, Level 7. Greenpeace activists had taken every protective measure possible for their personal safety and they were heroes, no doubt about it. My heart went out to them with gratitude the day that CNN reported their findings.

This elation was cut short however by another CNN story about two young reporters who traveled to the gates of the crippled reactors. They had heard that the general public had been notified that TEPCO would close the area in two days forever. (As in literally forever if such a thing is possible, due to the fact that the half-lives of some of the radioisotopes will remain for tens of thousands of years.) The young men drove to the gates of Fukushima Daiichi in street clothes, unprotected, and tried to communicate with the workers at the gate who were completely covered in protective gear and masks. "All they would do was quickly motion for us to turn around," and then "after 30 minutes we figured it was time to leave, especially when considering that we received as much exposure the day before when we were there." "Why did you go there?" the anchorperson asked. "Because that's where the story is," he replied matter of fact. I was shocked, then angry and very sad that these young men had put themselves in such grave danger. And we didn't know then but TEPCO had calculated the radioactive release to be far less than what it actually was. It wouldn't be for a few more months until the public was able to find out the real facts about Fukushima, and until then a lack of information only served to increase the frustration about the apparent disinterest with our own government and the media.

Never trust a computer you can't throw out a window.
– Steve Wozniak

Then came the day out of pure frustration when I slammed my laptop shut to break the silence a bit too loudly, apparently there was no open window around. One week and one new hard drive later, I was back at it but I was feeling jaded and completely disillusioned. Every day the news got worse, the kind that you find on the Internet from reliable sources, but that is also being completely ignored by newspapers and television. What should Americans do to be prepared? How could we protect our families? Enter again Food and Water Watch, remember them, from when I wrote the petition? Food and Water Watch was formed by Public Citizen, Ralph Nader's watchdog group, and when it became obvious that food safety required a lot of oversight, FWW was born. Watching Senator Boxer's growing anger directed at the NRC on TV seemed like a glimmer of hope so I picked up the phone and called them. I remember that my son was particularly restless that day and jumped on my lap with glee just as my call went through. FWW's Western Regional Director, Elanor Starmer was in town that day, which appeared to be the first break I had since the disaster began. She was not only understanding of the raucous noise in the background, but also appreciative of the call. *I was thrilled!* Would they want to schedule Senate meetings on this, I asked? "Sure, why don't we do that, and one with Senator Feinstein too." It didn't take long to be impressed by her thoughtful, professional manner. Then she dropped a bombshell. She just learned that the Environmental Protection Agency was scaling back random sampling and monitoring of rainwater and milk from weekly, then to biweekly and now to quarterly. "As in four times a year?" "Four times a year." *The largest nuclear disaster in world history*, and the only monitoring to our knowledge, was of rainwater, groundwater, and milk and *only four times per year.* Something was definitely fishy.

And speaking of seafood, the FDA had already announced that they did not plan to test north Pacific seafood for radioactivity whatsoever, even as TEPCO was releasing millions of gallons of plutonium-laced water into the sea as an attempt to cool the reactors. I mentioned to Elanor that some of the groups have been fortunate to work in coalition in the past to block the proposed Ward Valley dump in Southern California. "Sure," she said, "Go ahead and put together some people for the meeting and keep me posted." She was working on a number of projects already, so I was extremely pleased that she would lend her expertise and 3,700 signed letters from Food and Water Watch's supporters asking for increased monitoring. I informed her that I had found language from their letter writing campaign and incorporated some of it into my petition. She didn't seem to mind, which was a relief! We had a partnership of sorts and it felt good. Finally, good news. Great news at that.

First they laugh at you, then they ignore you, then they fight you,

then you win.

– Gandhi

So I took to my trusty laptop to reach out to others who might be interested in the meeting. In some ways what was happening in Japan reminded me of working to block radioactive waste dumps in the U.S.: in this case, however, the entire country was turning into a dump via air particulates from one radioactive origin (Fukushima Daiichi) colliding with a late and long rainy season in California. According to numerous reports on-line, several radiation detectors at existing nuclear power plants throughout the U.S. were picking up readings of radioisotopes identified as coming from Fukushima Daiichi. Physicians for Social Responsibility, or PSR, was of the many groups, along with Greenpeace, who had opposed plans for Ward Valley back when we were fighting the dump. US Ecology, the company licensed to dig the dump, had been cited in every state in which they had operated, even leaving the keys in the gate at Sheffield, Illinois, when it was discovered that radioactive tritium had migrated to tree leaves hundreds of yards offsite. Several years later the dump had been defeated after motivated citizens made it abundantly clear that California taxpayers would be expected to pick up the tab just as taxpayers in all the other states had been. It didn't help matters (or did, depending on how you look at it) that the land for the dump was also habitat for the endangered desert tortoise, an animal that is, ironically, millions of years old. Ironic because if radioactive waste was dumped there, it too would have remained for millions of years and would most definitely have destroyed a species which had managed to survive and thrive much longer than most. Another crucial reason to defend Ward Valley is that it was also ancestral land for the Shoshone and Fort Mojave Indians. A coalition of tribes formed and was crucial in helping to defeat the dump, and I was very proud to say that they trusted us to be part of the coalition. Shortly afterward I moved to Northern California to a small town to continue working in coalition to block the dump. Part of that work entailed traveling to Washington,

24

DC, to lobby on behalf of our coalition to block bills containing language mandating the Ward Valley ditches.

Then there was the Yankee Rowe campaign, which proved once and for all to the public the real purpose for at least some of the radioactive waste they were built to receive. At Rowe, Massachusetts, a nuclear reactor was being unceremoniously decommissioned and its radioactive components classified "low level" were readied to be hauled via 18-wheelers down Interstate 95, the major East coast corridor, and then dumped in at Barnwell, South Carolina. By now I was working with Nuclear Democracy Network (a group committed to working on safe energy issues for nearly 30 years and now the Ecological Options Network). With their help I initiated a campaign involving Greenpeace and several small local Massachusetts groups for an educational bus tour to Barnwell. We followed the trucks carrying the reactors at a relatively safe distance and armed with Geiger counters. Even while keeping back behind the trucks at a relatively safe distance, we were never comfortable with the exposures we were getting.

It was a relief when our efforts brought national attention to the fact that the Yankee Rowe decommissioned reactor was being buried in a dump similar to the one planned for Ward Valley. I still remember opening the New York Times A Section to the large photo that accompanied our action: we staged a mock cemetery scene, dressed in radiation protection suits, many of us lying on the ground in protest. Our action was meaningful in that up to this point we had only heard about claims from the nuclear lobby that Ward Valley was intended for "harmless" booties and gloves from nuclear medicine. At Barnwell people finally learned the truth: Ward Valley was intended for decommissioned nuclear reactors categorized "low-level" radioactive waste. The same category as those booties and gloves, however with significantly higher levels of radioactivity involved. It took several days to make the bus trip down I-95 and we met many motivated and concerned parents along the way. These were people who opened up their homes to our exhausted, yet exhilarated crew, strangers on the one hand but with whom we felt a kinship that transpired that minor detail. One breakfast that we shared, after 15 hours of driving, felt and still feels like one of the very best that I ever had. We felt we were making an important contribution and there is a real feeling of satisfaction which goes along with such undertakings. Mighty forces came to our aid.

During both campaigns I became an admirer of Dr. Helen Caldicott, a medical doctor ahead of her time (as visionaries will be), who was staking her professional reputation on the fact that there were no safe levels of radiation, whatsoever, from nuclear power production. None. Physicians for Social Responsibility (PSR), Nuclear Information and Resource Service (NIRS), Union of Concerned Scientists and more experts also strongly believe that low-level radiation causes cancer. Their knowledge inspired a generation of anti-nuclear and safe energy activists, and currently a new generation is emerging. A growing body of experts now believes that chronic, continuous low-level radioactive exposures are even more dangerous than short, strong bursts of radiation. The very type of chronic low-level releases being witnessed now from Fukushima Daiichi. Luckily, some of the experts from these groups were willing to participate in the upcoming Senate meetings. I was and remain indebted to them for helping an isolated and anxious mother such as myself.

10

How Safe is My Kitchen?

As mentioned earlier, California was experiencing a prolonged rainy season in 2011, well into June, three months after Fukushima Daiichi erupted, AF. This meant that U.S. crops were being doused with radioactive cesium, strontium and iodine (and there are, literally, hundreds more). A few of these radioisotopes were being reported by the UC Berkeley's School of Nuclear Engineering (UCBSNE), where students and faculty were testing a variety of rainwater from their class rooftop in Berkeley as well as random sampling of creek water, topsoil, and a variety of dairy and produce from around the region. Cesium had been detected in organic cow's milk in Sonoma, California. Radioactive iodine was detected in spinach and lettuce. And experts were predicting that root vegetables could become more contaminated due to their prolonged contact to potentially radioactive soil. "So, let me get this straight," I told myself, "No milk, no french fries, and no mac and cheese." The very foods that just about cover most young children's diets especially if they are picky eaters, and most are picky eaters in their younger years.

Now the most important question for me seemed to be "what was no longer contaminated by radioactive fallout?" And how did topsoil contaminated by these deadly radioisotopes affect agricultural standards and regulations. It was extremely difficult to wrap my brain around that question. Every principle I held to be true in my studies and practice as a holistic nutritionist was now being held hostage by the shock of radioactive contamination due to the tragedy in Japan. Tap water was definitely out of the question, as was the over-the-counter water pitcher filter we had been using. It took weeks to search for bottled water dated pre mid-March 2011, while simultaneously looking for an affordable water filtration system. In addition to stocking up on frozen berries, mushrooms, and other vegetables from the 2010 harvest, I created an extra pantry for tetra packs of miso soup, rice, hemp and almond milks and an assortment of canned beans, legumes and vegetables. Bags of rice, beans and legumes were also bought in bulk. For once it didn't seem to matter quite as much if they were organic, because buying that in bulk would have eventually been cost prohibitive. In any case maybe this was just a false sense of security, but at least at that moment in time "organic" didn't seem as important. I took things one day at a time as I moved around the kitchen in a bit of a trance. Every protective instinct I had as a mother seemed amplified, because I knew that my son was much more vulnerable to radiation in food than my husband and I. Just relatively safe compared to *radiated* could suffice just fine, and so potatoes were off the shopping list for the immediate future. Maybe the months immediately following the explosions and first plumes were the worst and prayers would be answered that food would somehow be safer next year. Hopefully the radiated fuel pools wouldn't explode and make the initial explosions seem mild in comparison.

Nuclear fallout is a harmful and mysterious tragedy that we can't see, taste, hear, smell or feel. Rather than recoil in fear from Fukushima Daiichi it really only served to empower me to further action. And it's okay to be afraid too, fear is not a weakness. If properly directed, fear and even anger can be powerfully motivating. This is what Ghandi said about anger, *"I have learned through bitter experience the one supreme lesson is to conserve my anger, and as heat conserved is transmitted into energy, so our anger controlled can be transmitted into a power that can move the world."* Denial is not a solution and it only enables the problem. If I ever feel fear I remember to ask myself, first, what it is really about, and then act on it, because fear itself is really a survival mechanism that can be transformed into action. What is needed especially now is action, educated and informed action. *Please note that in the following pages and in the afterword are very real tangible and effective steps for you to take to make change happen.*

> ***Out beyond right and wrong there is a field, I'll meet you there.***
> – Rumi

Our little boy played indoors during most of the spring rain, or was carefully covered by his big hood whenever we ventured outside. Jumping in puddles? Absolutely out of the question, even in high rain boots. *"Silent Spring on steroids,"* I grimaced. The ghost of Rachel Carson nudged again. Then I learned about a radiation protection workshop given by a local holistic health practitioner. A dear friend, (another Rachel, a long-time nuclear activist and artist that I greatly admire) told me about the workshop and how she came away feeling empowered and optimistic. She had heard about a type of water pitcher which, when used with drops of a special formulated version of zeolite, would help to improve water quality. Zeolite captures chemical toxins and heavy metals in a unique molecular cage and then removes them from the body via bodily excretions. The celebrated doctor Gabriel Cousens calls zeolite's detoxifying benefits "miraculous." Still, we reminded ourselves that even something like zeolite isn't a panacea. Nothing known to humankind can completely rid our environment of excessive radiation, and uranium should remain as nature intended, buried deep underground. But I'm open to being hopeful that it may be possible to minimize risk by taking some protective measures. The real solution overall is to end nuclear power as soon as possible. There was news by early July 2011 that Germany and Switzerland were planning to transition from nuclear power, due to the accident at Fukushima Daiichi, with Italy resolving to do so shortly after. Again, good news. If in the meantime continuing to using caution wherever possible when choosing foods and taking supplements that I believe will help decrease exposure to my family will decrease my concerns, well then I'll stay open to exploring possibilities.

Consider the amazing and true story of Dr. Tatsuichiro Akizuki, director of the Department of Internal Medicine at St. Francis Hospital in Nagasaki. Dr. Akizuki quite literally *prevented* radiation sickness at the hospital after the atomic bomb was dropped there in 1945. He accomplished this by using select foods. That's right, food. He instructed cooks at the hospital to prepare a strict macrobiotic diet for his staff and patients that was also completely devoid of sugar. No one in his hospital succumbed to radiation sickness, even while survivors farther away died from radiation exposure. Miso soup, unrefined brown rice, and seaweed such as kombu and nori are the cornerstones of the macrobiobic diet. The concept of food as medicine has never proved more powerful.

The new water pitcher and filters arrived one week after the zeolite drops, which provided some semblance of hope. But I was skeptical, even amid stories of children around Chernobyl being fed zeolite cookies to fight radiation sickness in the 80s. Then I read that powdered zeolite had been used at the Chernobyl site along with boron to lower radiation levels, at Three Mile Island in Harrisburg, Pennsylvania, and at the Hanford reactor in Washington. Upon checking further it appeared affective and non-toxic, the only side effect being possible dehydration during detoxification so it is necessary to stay well hydrated while using zeolite.

The Standard American Diet (SAD) has, for many of us, become the norm. The so-called "nuclear family" tends to focus on convenience foods with low nutritional value (think fast food, microwaveable meals, processed snacks). So, while nutritional supplements shouldn't replace healthy food by any means, in the Atomic Age it seems to be more important then ever to shore up deficiencies by taking a good high-quality multivitamin and mineral supplement. Iodine, calcium and magnesium have become more important than ever. I learned that radioactive strontium, an extremely dangerous radioisotope and byproduct of nuclear power and accidents, is absorbed by the body much like the mineral calcium, and if we are lacking calcium already, strontium uptake will occur more quickly. Thyroid cancer is caused by iodine-131, however if a person already has sufficient iodine in their system, it will be less likely to attract iodine from the environment. Vitamin C is a phenomenal anti-oxidant long recognized to protect the immune system and is one of the most valuable nutritional supplements available to us. It was the subject of a large body of research by renowned scientist and Nobel Prize Laureate Linus Pauling (one of only a few people to have won two unshared Nobel Prizes, one in Chemistry, and the Nobel Peace Prize). From the kitchen table I continued to research supplements whenever my son would nap, because I was feeling desperate to figure out a way to protect him.

Kelp, chlorella, blue-green algae and spirulina all help to maintain proper pH levels and assist in cleansing and detoxification. There are numerous studies that indicate they help to protect cancer patients from the negative effects of radiation treatments. Rosemary extract contains carnosol that has been proven to have anti-carcinogenic and inflammatory properties specifically for mammary tumors. Agari Gold mushroom capsules, and Eleuthro ginseng have also been studied with positive results to help fight radiation sickness and can all be ordered on-line or in store. Seaweeds need to be properly sourced and those sources will become few and far between as Fukushima Daiichi continues to dump plutonium-laced seawater into the ocean. Carraneegan is used in everything from organic strawberry milk to baked goods to nutritional supplements, yet it is being used for all without proper testing. Mushrooms have long been known to hold valuable medicinal properties, and they too are under threat because they absorb nutrients via both the soil and the atmosphere. Eggs, considered to be the richest source of concentrated animal protein, are now also suspect. They are like seeds and concentrate not only nutrients (which is why organic eggs are considered to be the most important protein staple in a ovo-vegetarian diet) but toxins as well. However all commercially and even free range raised animals coming in contact with radioactive fallout will concentrate radionuclide toxins in their tissues, which are then passed up the food chain to the humans who consume them. Consequently eating low on the food chain is more important now than ever before.

So Long, "Salad Days"

Gone were the days of innocently preparing a meal for my family. I longed for the times I would fuss over not finding organic eggs or grass-fed beef. The realization that commercial grain-fed beef might actually be healthier now than grass fed, which has always been considered superior due to its high omega 3 content, could now actually be more contaminated by cesium with a half-life of 30 years and which remains hazardous for 600 years. And fish, well, how much longer would I even consider feeding my son salmon or any other seafood for that matter. Fish oil supplements had been a mainstay in my health regimen for many years, but for how much longer would fish oil be safe, or some of other nutritional and herbal supplements for that matter? Carrageenan is common in supplements and many food products and baked goods. It is sourced from red seaweed that is sourced from, of course, the sea. Without monitoring and testing, how will be know which seaweed is safe to allow into our diets?

Suffice to say that just standing in the kitchen prepping a meal had become nothing short of mind-boggling, not to mention frightening. What the heck was safe to eat and drink anymore? *From a nutritionist's perspective, crazy making. From a mom's perspective, worrisome. From a former radioactive waste activist's perspective: scary as hell.* The three-headed beast was born. I would ask myself, "How will this affect organic regulations?" As far as I could tell, Genetically Modified Organisms (GMOs) and irradiation were the most important threat to organic standards. After much thought I've decided to stop using organic butter for the time being and instead use coconut butter and oil (which has added benefit of supporting healthy thyroid function). However, oils in general concentrate toxins so this alternative will only be relatively "safe" temporarily. Taking a high-quality multivitamin including iodine is essential, because the thyroid will seek out iodine -- even radioactive iodine-131 -- if the healthy kind is unavailable. My plan was to continue to buy products from companies held to the highest standards and hope that they would soon learn about the Fukushima threat and source their ingredients accordingly. Clearly, more petitions will be needed. Citizen action is essential to our survival.

31

11

RETHINKING REALITY

Reducing exposures in general seems like the only course of action worth taking. Renowned nuclear expert turned watchdog Arnie Gundersen from Fairewinds has said that the indoors becomes more radioactive than outdoors due to concentrations of contaminants accumulating inside. He recommends — for those of us living in Japan and on the West coast — to keep windows closed during summer, change air conditioning filters in homes and cars regularly and continuously. To avoid spreading possible particulates, he recommends wet dusting and mopping, as opposed to dry dusting; we should also use High-Efficiency Particulate Air or HEPA floor filters. Taking shoes off at the door helps to avoid bringing the outdoors inside (as the Japanese have done for centuries). For water, there are reverse osmosis and carbon filtration water systems available and using a combination of both has been said to be the best way to purify water, although it is widely debatable how much radiation such filtration will actually remove. Again, without testing, how do we know? Reverse osmosis is fine under normal circumstances for the short term, however RO water has been de-mineralized, similarly to distilling water. Which means that drinking either indefinitely would deplete valuable minerals from the body. Taking a multivitamin and mineral supplement would help to replenish stores. And, while carbon filters are fine for bathroom sinks, bathtubs and showers, they are not generally regarded as the best for drinking water.

And then there are Epsom salt baths popularized by naturopath Dr. Hazel Parcells, author of *Live Better Longer* and known as the "Grande Dame of Alternative Medicine." She held a medical degree in chiropractic, as well as two PhDs (one in nutrition, the other in religion). Ms. Parcells lived to be 106 years young in New Mexico and was particularly concerned with radiation due to the bomb blasts of the 50s and 60s. She is known to have guided many people through detoxes who had previously been exposed to radiation. If something as simple as an Epsom salt bath could help guard against Fukushima, maybe there was hope. I resolved then and there to start a website and reach out to others, especially people in Japan, who were dealing with the horrible situation. I resolved to stay hopeful that there would be ways for our family to stay healthy *AF.* What choice did I have?

Nuclear Disasters and Lessons Learned

Facts are facts. There have been three major nuclear disasters to date: Three Mile Island in 1979, Chernobyl in 1986 and now Fukushima Daiichi in 2011. There are many more, smaller nuclear accidents and near misses every year. But this is not considering the problem that exists in between small and large events. Fukushima Daiichi serves to highlight the fact that there are 104 nuclear reactors in the U.S., many of them nearly 50 years old and approaching re-licensure. The General Electric Mark I reactor, the type at Fukushima, is also prevalent in the U.S. and radioactive spent fuel pools are on all of these sites. The NRC expanded the evacuation zone near Fukushima Daiichi for Americans from 10 miles to 50, however one must wonder how it would be possible to adhere to that guideline here in the U.S. given concentrations of our population and proximity to reactors in the Northeast and Southern California, to name two regions. Most reactors in the U.S. are in heavily populated areas. A third of the population in the U.S., or roughly 120 million people, live within a 50-mile radius of a nuclear reactor. Hundreds of accidents at these plants occur annually. When do we say "enough?" When will Atoms for Peace finally end? They certainly weren't thinking peace of mind.

From time to time I seriously considered moving my family farther away from Fukushima Daiichi, maybe back to the East coast of the U.S. where I grew up. Then came the news that the U.S. government approved plans to accept radioactive waste from Germany's nuclear program and to incinerate "low-level" nuclear waste at a Department of Energy facility in Oak Ridge, Tennessee! This could just be the beginning. And with this came the growing realization that all of our 104 nuclear reactors in the U.S. emit radiation as part of their function, adding more radiation to an increasingly radiated world.

One day someone posed an interesting analogy during a phone conversation: picture frogs in a pot of slowly heating water. The water heats to boiling so slowly that the frogs don't realize what's happening, and never do, they simply die. It seems like the human race is nearing the boiling point where nuclear science is concerned. French researchers confirmed in a report in early 2012 that childhood leukemia rates are greatly elevated among children living near nuclear power plants. The *International Journal of Cancer* published a study in January of 2012 titled, "Childhood Leukemia Around French Nuclear Power Plants — the Geocap Study 2002-2007." One could then rationally wonder if a future study won't find that the food grown near nuclear reactors isn't somehow affected as well, and the water that feeds the crops. "So you see," I would tell myself, "There really is no place safe anymore where nukes are concerned." The only hope we have is to remember events and to act on them, to not become the frogs in the boiling cauldron… We can do better for our kids.

One of the lessons of Chernobyl is that many of the foods we know to ordinarily be the healthiest are in fact those which attract some of the highest levels of radiation from nuclear fallout: seaweed, obviously, and especially in the North Pacific where it is at risk for radiation contamination. Mushrooms, spinach, kale, lettuces, berries of all kinds, cheeses, and fresh cow and goat milk are also at risk because we know from Chernobyl that they attract some of the highest levels of radiation. Mushrooms, for instance, absorb nutrients through not only the soil, but primarily the atmosphere. UCBSNE, as mentioned earlier, is one of the only organizations which sampled and publicly posted findings of an array of produce, along with milk and rainwater which have been found beginning in late March 2011 to contain levels of radioactive iodine, cesium and strontium. All are listed with the caveat that levels are "to be expected," "within safe ranges" and "comparable to taking a cross-country flight."

At the root of it all is the question "at what level is radiation from nuclear accidents harmful?" Herein is where much dissention exits between the nuclear industry versus a growing body of scientists, physicians and nuclear experts such as Dr. Caldicott, Physicians for Social Responsibility, Union of Concerned Scientists, Fairewinds and many more. The former includes the Nuclear Regulatory Commission, International Atomic Energy Commission and the powerfully enmeshed and entrenched nuclear industry (dating back to Atoms for Peace). They state that all levels found are "to be expected." Expected according to what? When have we had an ongoing nuclear disaster with releases extending for an unknown period of time? Is Level 7 on the International Nuclear Event Scale even relevant any more? Will the level system be raised to 8 or higher now that we know these disasters can roil for months, possibly years? Scientists in the know emphatically state that there is no such thing as safe low-level radiation from nuclear power. Period. In 2006 National Academies of Sciences stated, *"There is a linear dose-response relationship between exposure to ionizing radiation and the development of solid cancers in humans. It is unlikely that there is a threshold below which cancers are not induced."* Clearly the time for monitoring, as stated in my petition, is now, as in yesterday.

12

Mighty Forces

Action May Not Always Bring Happiness,
But There Can Be No Happiness Without Action.
– Gandhi

I did something pretty unusual for our family and arranged child care for June 2 and 8, the dates set for meetings with Senator Boxer and Feinstein's aides, respectively. For the first time I could appreciate how parents who work outside the home entrust their children for several hours to someone else. Luckily these were people we knew well and it would be possible for me to concentrate on the task at hand: traveling 30 miles by cabs and train in rush hour to make the meeting in downtown San Francisco in time to organize our group beforehand. My son got to see mommy all dressed up and carrying a briefcase. He wanted to go to the meeting too, but once I explained that there wouldn't be any toys his enthusiasm waned considerably. "All in good time, son," I thought. "Plenty of battles to come for your future, too, unfortunately." A somber thought. On the one hand, how fortunate to be a part of something that might impact an important issue in the larger, positive sense. On the other I couldn't help but grumble a bit, "Why isn't our government doing this already?"

It was a blessing all the same, and this is when I realized that the mighty forces had indeed arrived! Leaders from Food and Water Watch, Citizens for Health, Physicians for Social Responsibility, Nuclear Information and Resource Service, Greenpeace and other individuals attended over both days. One gentleman, representing as a citizen and grandfather, drove several hours to say that he was planning on starting up his own radiation monitoring center if the government wasn't going to test food and water. A concerned housewife, still breastfeeding her youngest child, bravely accompanied our group and contributed much to the discussion. I couldn't imagine how this all seemed to her. This was all somewhat old hat for me and yet still anxiety provoking, but what must she be feeling, knowing that her breast milk could be more at risk than ever? And then there was the dedicated nutrition student who wanted to know what was safe to tell people to eat now that so much of our food supply was at risk, and that until they monitor, how will we know its safe to eat? There were many questions, and we tried to achieve a mood that was one of collaboration: how could the Senators, also being Californians, help us keep our children safe? Food and Water Watch presented their letters along with my petition and together there were close to 6,000 signatures. *No small potatoes, for sure.* Besides the top regional Senate aides in the room, we also had a representative on speakerphone to Senator Feinstein's DC office, which likely increased our chances of the Senator hearing our concerns.

In addition to the "asks" in the petition, we asked for full congressional hearings into the matter. There were too many questions to be answered, so many in fact that it seemed like a quagmire. Senator Boxer is the Chairwoman of the Committee on Environment and Public Works (EPW), so she is perfectly positioned to hold such hearings; and Senator Feinstein chairs the Appropriations Committee with Food and Drug Administration (FDA) oversight — so we asked that they hold hearings and provide adequate funding for food safety inspections by United States Department of Agriculture and FDA. If the radiation isn't there, we need to know. *"Prove us wrong."* And one important observation is that it flies in the face of reason to think that between the Environmental Protection Agency (EPA), FDA, National Oceanic and Atmospheric Association (NOAA), and the U.S. military — that among all these powerful groups only EPA is testing water and milk samples, and merely four times a year at that? Is it even necessary to ask for funding? We asked the Senators to write letters to EPA, FDA and NOAA directing them to get on the same page and to release findings they most likely already have collected. Even if the news is bad, we as citizens have a right to know. Again, prove us wrong. As it turns out, asking for a Freedom of Information Act request (FOIA) could take months to procure, with results very likely blacked out and impossible to discern, so instead we asked for reports from the General Accounting Office (GAO) and Congressional Research Service (CRS) which would help to provide a foundation for the hearings.

One notable highlight of the Boxer meeting was when Dr. Robert Gould with Physicians for Social Responsibility raised the issue of the Senator proposing an immediate phase out of nuclear power such as the Germans and Swiss were doing (and the Italians are now joining them). Given what we now know about wind and solar alternatives to nuclear and coal gaining in momentum, this would seem logical and necessary. Diane D'Arrigo from Nuclear Information and Resource Service has been an expert on radioactive waste issues for over 20 years. Working tirelessly to protect the public, she worked to block the Ward Valley dump and others back in the 90s and to this day. She was on hand via speakerphone from DC and made some powerful statements, including telling the Senate staff that there are *hundreds* of radioisotopes that are byproducts of nuclear power production, and some of them hazardous for a hundreds of thousands of years or MORE. Although once again, the only one that ever seems to be mentioned in the media was iodine-131, with a half-life of eight days.

Over the next several months we held more meetings, and our group (initially myself and Diane D'Arrigo) grew to include Jim Turner, Chairman of Citizens for Health, Mali Lightfoot, Executive Director of the Helen Caldicott Foundation, Jim Heddle, Marybeth Brangan and Rachel Johnson of Ecological Options Network, and Cindy Folkers of Beyond Nuclear. Collectively their expertise and experience working on nuclear issues was immense and I will always feel a huge sense of gratitude for their knowledge and support. So what really came out of the meetings? All is yet to be revealed. On one hand we felt heard and supported, on another, we are up against the nuclear industry and their hugely influential lobby on Capitol Hill. There was a sense that our representatives want to help us, yet on the other hand the nuclear behemoth is huge and deeply embedded in our culture going back to the atom bomb. Nonetheless, we persevere. I felt truly and deeply honored to be in the company of so many smart people while launching our campaign for food and water monitoring. We continued to work together as a fledgling coalition of groups – name still to be determined – firmly committed to strategizing the next step.

It is really important to remember that part of the strategy should be not only closed-door meetings but also a community-driven effort. Direct action has roots going back to 1849 when abolitionist Henry David Thoreau wrote his essay, *On Civil Disobedience,* which explored standing up for what you believe in, basically. The Suffragette, Civil Rights and Environmental movements have all grown from the seeds planted by Thoreau. Most recently the Occupy Movement has reminded us that ordinary citizens can have an extraordinary impact on our culture. We do not always have to have the answers to make change happen. The mighty forces will arrive. This little book is intended to help the average citizen remember that we all need to occupy our own position within society in order to provide a future for our kids. Healthy debate is supposed to be central to our Democratic principles. Think of those who have gone before us who have given their lives in fighting for our rights to free speech. Women in the United States fought for 70 years until we were granted the right to vote in 1920. Yet all too often it seems we forget the power that we do indeed own. If we remain silent, we have no power whatsoever. And even for those who may feel their vote doesn't count, there is always community volunteerism and a number of other ways to get involved and make change happen. But certainly we need to hold our elected officials accountable. We need to be accountable ourselves for who we allow in office in the first place.

13

FUKUSHIMA: WHAT REALLY HAPPENED

The Facts Revealed

By the late summer of 2011 the on-line world community pieced a more complete picture of the events at Fukushima Daiichi. They went like this:

The 9.0 East Japan Earthquake halted all function at the four GE Mark I Boiling Water reactors. The irradiated fuel pools then began their hellish descent into uncontrolled criticality. Each Mark I reactor contained 100 tons of fuel. Unit 1 suffered a total meltdown. Hydrogen gas built up and the containment building exploded. The images of the live surveillance video looks like a nuclear bomb detonation. The reactor core and eight years of spent fuel previously stored onsite melted down, and then began to enter the atmosphere unabated.

Two days later, Unit 3 suffered a hydrogen explosion as well. This differed, however, from the first explosion in that this was weapons grade plutonium used as fuel. Hydrogen explosion triggered a nuclear explosion, ejecting the contents of Unit 3 into the atmosphere.

On August 1, 2011, Professor Tatsuhiko Kodama, head of the Radioisotope Center at University of Tokyo, testified before the Committee on Welfare and Labor in Japan. He unequivocally stated: *"When we research the radiation injury/sickness, we look at the total amount of radioactive materials. But there is no definite report from TEPCO or the Japanese government as to exactly how much radioactive materials have been released from Fukushima."* So, he continued: *"Using our knowledge base at the Radioisotope Center, we calculated. Based on the thermal output, it is 29.6 times the amount released by the nuclear bomb dropped on Hiroshima. In uranium equivalent, it is 20 Hiroshima bombs. What is more frightening,"* he continued, *"is that whereas the radiation from a nuclear bomb will decrease to one-thousandth in one year, the radiation from a nuclear power plant will only decrease to one-tenth."* This is important information that governments cannot block anymore thanks to the Internet. Why not be a part of history yourself and use it to help protect your family and future generations? Several months later, Frontline interviewed a former nuclear industry official in Japan who said that the Japanese had been "deluded" by the promise of nuclear power. Now, their nation will never be the same. It will take generations, if ever, for people to return to Fukushima and the surrounding region. It will take generations for the ocean to recover from the dumping of radioactive waste. Millions of people depend on the ocean for sustenance, an ocean that is now polluted by radioactive fallout.

If one wants to go fast, go alone,

if one wants to go far, go together.

– African Proverb

People these days have it so much easier than 20 years ago before the birth of the Internet. The differences are tremendous, and for myself this experience has possessed a time-traveler quality. Back then there were reams and piles of files and paper, desks loaded with index cards, file cabinets stuffed to capacity. Now in cyber world communication means not having to wait for returned answering machine calls. Ideas and information flow at lightning speed to broad audiences. New information is always emerging and capable of being transmitted worldwide at our fingertips. Libraries have computers for those who don't own them ... in other words, no excuses any longer to not take action.

One valuable trick recently learned is that it's extremely easy to set up conference calls for a large group of people just by searching on-line for a free phone conferencing service. Group e-mails are a cinch, and Facebook and Twitter help to deliver messages to the masses at lightning speed. There are even applications and websites that help organize large numbers of people for meetings.

As of this writing there have been three nuclear meltdowns at Fukukshima Daiichi. Three. And there is still no complete containment of radiation. So even as the long hot summer burned, we continued the good fight. A small group of highly motivated citizens in Northern California are beginning to organize a system for independently monitoring the radioactive fallout from Fukushima Daiichi in air, food and water.

14

FUKUSHIMA SOLIDARITY — A GROWING MOVEMENT

The People, United, Will Never Be Defeated.
– rally chant

Perhaps one of the biggest breakthroughs happened in October when one of our members, Mary Beth Brangan, who was planning to be in DC, offered to meet with our Senators' staff there. This would be the third set of meetings to date. I stayed in California to care for my son, but both Senators' aides had kindly arranged a conference call dial-in number so I could take part. Several of our group who work in DC agreed to attend the meetings, so I set them up by email and phone while going about my usual day as a mommy. Our elected officials are incredibly busy people and their staff welcome input from the public, so never underestimate the power of picking up your phone.

The topic, as usual, was our coalition's concerns that radioisotopes such as cesium were still being detected by University California Berkeley School of Nuclear Engineers (UCBSNE) researchers in cow's milk sampled from Bay Area shelves, and that nothing was being done to alert the public. But really, it was the topsoil we were (and still are) the most concerned about, because if it's in the soil obviously the problem is widespread. In fact, by October, seven months after the initial nuclear disaster, UCBSNE were still finding radioactive cesium, strontium, and more in berries, lettuce, spinach, and a host of other produce sampled from the region. I learned after I hung up from the meeting that our group actually ran into Senator Feinstein in her hallway as they were leaving that day. By first-hand accounts, she appeared "shaken" when told about the cesium. "Oh no, I've been assured everything is fine," she said. "No, Senator, unfortunately it isn't fine," replied our team. The Senator gave us her private fax number and asked that we send her our information marked "personal and confidential." We did, and as of this writing, still no word from her office about her thoughts on this critical matter. More silence. One could get used to it if not careful, but our group was determined to not be complacent.

By Halloween of 2011 there was much more to be concerned about than the costume my little boy would be wearing. Mr. Gundersen's Fairewinds website posted the following announcement:

Washington, DC – Oct. 31, 2011 – *Today Scientist Marco Kaltofen of Worcester Polytechnic Institute (WPI) presented his analysis of radioactive isotopic releases from the Fukushima accidents at the annual meeting of the American Public Health Association (APHA). Mr. Kaltofen's analysis confirms the detection of hot particles in the U.S. and the extensive airborne and ground contamination in northern Japan due to the four nuclear power plant accidents at TEPCO's Fukushima reactors. Fairewinds believes that this is a personal health issue in Japan and a public health issue in the United States and Canada.*

Arnie Gundersen at Fairewinds reported Mr. Kaltofen's findings on the Fairewinds.com website. As you know by now, the media has been disregarding the dangers and realities of Fukushima Daiichi, so I want to share Fairewinds information with you here.

"To summarize the paper, citizens, some doctors and scientists, some bloggers, some farmers, around the world provided samples to Mr. Kaltofen who analyzed them for Fukushima radiation. An example of what he found is a slide that contains air filters from cars in Japan and in the United States. Cars in the United States hardly have any radiation in their air filters. Cars in Tokyo had quite a lot, way too much. Cars in Fukushima Prefecture were incredibly radioactive. Now I think it is important because the nuclear industry will say, well everything is radioactive and therefore we should not worry. Well, the Seattle data show that not everything is radioactive. And it shows that the people in Japan received enormous exposures of particles into their lungs and into their digestive systems, during the course of the accident.

Another piece of information is that Fairewinds viewers were able to send in children's shoes from Japan. Mr. Kaltofen has data that clearly show that the concentration of cesium on the kid's shoelaces was astronomically high, around 80 disintegrations per second. What does that mean? Kids tie their shoes, their hands get radioactive and it goes into their G.I. tract. If it is on the ground, it is in the dust in the playground and it is in their lungs. I think that between the two, the air filters and the children's shoes, it shows that there is a severe personal health problem in Japan that will manifest itself in cancers over the next 10 or 20 years.

Kaltofen did not just look at Japan. He set up monitoring stations in the United States as well. Two of the three monitoring stations in the United States did show hot particles in the air in April (2011). Since then, there have not been any hot particles. But in April, it is clear that, at the worst of the accident, hot particles were wafted across the Pacific and deposited in Seattle and in Boston at least. There is also data that indicate contamination on the ground in the Cascades, which are a mountain range that are right up against the Pacific Ocean."

My heart goes out with gratitude to Fairewinds for sharing this invaluable information, another example of "bearing witness." We certainly aren't getting this kind of information in the mainstream media or from governmental agencies.

In early November 2011 I learned about another petition started by a mother of a young child. This woman was Japanese, living in the U.S., and her name was Tomoi. Most of her family lived back in Japan and she was becoming more and more distressed regarding reports of the Japanese government's decision to burn millions of tons of rubble, a good deal of it radioactive. The Japanese government had resolved in June to burning the rubble rather than monitoring it for radioactivity. Japan incinerates municipal waste on a regular basis, and every city has at least one major incinerator. Clearly Japan was overwhelmed by the sheer magnitude of the threefold disaster and officials had decided that citizens should "share the pain" by distributing the rubble around the country before burning it and then dumping the ash in Tokyo Bay and other locations.

Tomoi decided to take action. She contacted people who could assist her, in this case Mali Lightfoot with Nuclear Free Planet. Her on-line petition was signed by people around the world, led by long-time humanitarian Bianca Jagger. Women in New York City, New Delhi, London, Munich, Osaka, to name a few, presented Tomoi's petition to Japanese consulates and embassies in conjunction with a major demonstration sit-in by mothers in Tokyo. All of this was created and organized via the Internet. By now, our group had been meeting somewhat regularly and we presented Tomoi's petition to Japan's San Francisco Consulate and the Japanese Embassy in Washington, DC. In San Francisco we sat across the table from the Consul, a dignified man who was visibly moved by our offer of condolences for the terrible events his country had endured. We were firm, however, in our resolve to convey that we could not stand for "sharing the pain."

We had heard reports that young school children were routinely shamed into eating food from the Fukushima prefecture that contained levels of radiation, levels that were unacceptable and a threat to their health. Tomoi and her friends and family were and are doing everything in their power to learn and share information; they are an inspiration. Even when faced with cultural forces that tell them that they should not speak up, that women should be submissive and "shut up and stop complaining" (stated by an official in Tokyo after we delivered the petitions,) they have the courage to do what is right to protect their families. By speaking up here in the U.S., we are helping not only ourselves but Japan as well. Putting pressure on our elected officials and holding them accountable sends a clear message to the people of Japan that we have not forgotten them.

Our group officially became Fukushima Fallout Awareness Network (FFAN). We issued a press release and contacted local news outlets. We managed to convince a local TV station to send a cameraman to the Japanese Consulate in San Francisco. For the first time we were able to state to the public on television that radioactive cesium had been found by University of California to be in topsoil and milk samples taken from supermarkets in the Bay Area. If Japan continued to incinerate, it would only make the matter far worse. We took the opportunity to draw the two issues together as one: incineration from Japan would potentially only increase the radioactive fallout to California produce and dairy. FFAN in DC delivered a letter that detailed findings to date by UCBSNE and it was delivered as a reminder to Senator Feinstein again, (her San Francisco office was in walking distance from the Consulate). The video of the SF Consulate event was posted on-line and garnered approximately 15,000 views in two months. We continue to stand in solidarity with the mothers of Japan going into the new year.

"Don't lose hope, when it gets darkest the stars come out." — Unknown

If you've read this far I'm hoping you will be inspired to action like Tomoi. Science has proven that we can draw a line between nuclear fallout and cancer. We do everything we think we can to protect our families but we aren't doing enough. When Fukushima Daiichi's radioactive fallout hit the West coast of the U.S. as estimated on March 18, my family was at an outdoor barbeque fundraiser in Monterey for our friend's baby son Matthew who had just been released from Stanford Hospital in recovery from leukemia. A few months before, a routine post delivery doctor visit found that "something didn't look right." This led to a medic helicopter whisking Matthew's mom, dazed and confused, toward Stanford with her infant son in her arms. It was only upon landing that she learned that his white cell count was over 30,000, where a normal count would have been 30 to 100. Mom and dad lived a nightmare most of us pray will never happen to our family but which we hear about more and more: week upon week meant more medical disclaimers to sign as little Matthew's battle for survival became a quagmire of experimental medical experiments, oxygen tents and sleepless nights. With only hours to spare his precious life had been saved, and now seven years of chemo are recommended as further treatment. So we were all in a celebratory mood that day, March 18, when our friends stood in a light drizzle listening to live music while Matthew's dad cooked hot dogs on a big outdoor grill.

My three-year-old son danced in the light rain as his hood kept falling back. He looked skyward to catch the mist on his face. The news didn't come for several weeks that the day we celebrated a cancer remission in the rain in Monterey, was the very day radioactive fallout from Fukushima hit the West coast of the U.S. When I look at pictures from that occasion it is a haunting memory: the timing, the irony and the questions. *"How much radioactive cesium fell on our children that day, or radioactive strontium, iodine or the hundreds of other fission byproducts from Fukushima Daiichi? How much landed in their drinks, food and clothing?"* *"Hopefully nothing, most likely the plume was still up in Alaska on that particular day before heading south to Vancouver, Seattle and Portland,"* I would tell myself that in retrospect. But it's cold comfort thinking that someone else's kid was getting doused with radiation instead of mine. One thing is for certain if you choose to recognize history: cases of cancer in all forms will increase after Fukushima, all across Japan and the western region of the U.S. and very likely farther inland. Europeans saw it with Chernobyl, and now we will see it with Japan's disaster — Japan, Canada, the U.S. ... radioactive fallout knows no bounds. If I ever think of giving up, I think about Matthew, especially because his leukemia came out of remission and as of this writing he is struggling for life while waiting a bone marrow transplant.

While young children, the elderly and immune deficient are at particular risk, Fukushima Daiichi will affect us all globally for generations to come just as at Chernobyl. One thing we do know is that we are at a crossroads with nuclear power. We can and should oppose the loan guarantees our President has promised the nuclear industry. We can and should reduce consumption and embrace sustainability. We can and should transition to renewables like solar, wind and others.

Our long hot summer will lead to fall and what then? What of our winter and all seasons to come? Do we simply wait for the next nuclear meltdown to happen? Or do we transition to renewables now. Again, I took to the Internet to quench what had become a big thirst for knowledge on the issue. It was the least I could do for my son, having brought him into this world and all.

15

Bridging Over

We've all heard the terms "reduce, reuse, recycle" but clearly that maxim falls short. We need to "re-energize." Luckily for us, this is not a new concept and much research exists on the topic, forged as early as the 1970s and led by a visionary named Amory Lovins. He sounded the alarm early on about widespread catastrophic damage due to climate change and the consequences of nuclear power production. He was widely ignored. In his Foreign Affairs paper published in 1976, he argued that the world needed not new energy supplies but more energy efficiency. However, it's clearly hard for corporations to cash in on reduction of use, so this concept has been an extremely hard sell in a capitalist society. But now with "green" renewable energy outpacing coal and nuclear production there is a distinct window of opportunity to "bridge" to renewable energy. And Fukushima Daiichi has sounded a different alarm and provided a distinct window for survival as a species.

The more I studied what was happening at Fukushima the more it became apparent that, in reality, it is not enough to fight the good fight. As climate change activists engage in their own campaigns (anti-coal, anti-gas, anti-fracking, anti-oil and of course anti-nuclear) maybe we can all remember that we are all FOR safe energy and a healthy and clean environment. We all want and require healthy and clean food and water. Remember, these needs are not separate, but closely intertwined. We are what we eat, and we are "the environment." We need to start setting an example by talking about energy efficiency and reducing consumption — arguing for transitioning to renewable energy is not NEARLY enough.

Harvey Wasserman is an inspiring leader in the anti-nuclear activist movement for many years. He was talking wind and solar power back when few were considering the benefits. His Nukefree.org blog has been keeping the public apprised of the nuclear power industry and much of the news is daunting. It doesn't begin or end with Fukushima. The NRC is allowing plans to move forward for a new reactor complex in Georgia. The approval is the first for a new project since 1978. What interesting timing. Here the world is still reeling from the ongoing consequences of Fukushima Daiichi, three nuclear meltdowns and counting, and the NRC and industry response is to build a new reactor in the United States! Mr. Wasserman's Nukefree.org blog states:

"Among other things, the Commission raised questions about whether the AP1000 can withstand earthquakes and other natural disasters. Even now the final plans are not entirely complete. Only two other U.S. reactors -- in neighboring South Carolina -- are even in the pre construction phase. As in Georgia, South Carolina consumers are being forced to pay for the reactors as they are being built. Should they not be completed, or suffer disaster once they are, the state's ratepayers will be on the hook."

> ***If you're not angry, you're not paying attention.***
> – a favorite bumper sticker

We all have our own history, our own unique resumes. One doesn't need experience working on nuclear issues to get up to speed and make change happen now. We now have a choice that won't come again in our lifetimes, or even that of our children's lifetimes. Picture a fork in the road where, on one side we see the next generation of nuclear reactors and the deadly waste, cancers and other illnesses they will create for millions of years — or on the other side picture wind, solar, geothermal and biomass which all are available, safe, sustainable energy sources. Environmentalists have been calling for this transition for over 40 years, and it's hopefully not too late. Never before have we had this unique window in time to make it happen. But collective denial won't work.

We must demand monitoring of our food and water, reduce consumption, switch to renewable energy and finally put an end to nuclear power. An impressive community of smart and savvy citizens has grown considerably on Facebook and Twitter. The information that people are sharing on-line is immense, however, there is still a great need to spread that information more widely to communities where people need to better understand the dangers of nuclear power and its effect on our food supply and health, especially our children's health. Not only that, but also about the urgency of transitioning to renewable and safe sources of energy. I humbly suggest that you start and sign petitions, get out and raise your voice, start a fundraiser for a favorite group working on this issue, contact your elected officials for meetings (I'm assured that they want to hear from regular citizens even more than they do experts), write letters to the editor, and use on-line tools like YouTube to research and share your concerns. Then call your local Greenpeace office or one of the other amazing organizations mentioned above for assistance if needed. Greenpeace Japan and Green Action are two great organizations in Japan as well.

In the famous words of poet Khalil Gibran, *"Our children come through us, but are not of us."* It was my decision to bring a child into this world. Much like paying the rent or mortgage for a home there is an obligation to pay for our time on this planet, and to provide a safe future for our kids. Even in this age of questioning all sorts of puppet authority and power, we are not ourselves powerless. In many ways we are more powerful than ever before. Another favorite quote, this by the late anthropologist Margaret Mead, sums it up for me, *"Never doubt that a small group of thoughtful, committed citizens can change the world, indeed, it is the only thing that ever has."*

✳

Kimberly's petition led to the formation of the Fukushima Fallout Awareness Network (FFAN), which is a growing coalition of groups dedicated to the truth about radioactive fallout due to Fukushima Daiichi. As of early 2012 core FFAN members include Nuclear Information and Resource Service, Beyond Nuclear, Citizens for Health, Ecological Options Network, Nuclear Free Planet, and concerned parents and grandparents. Other groups present at FFAN-organized Senate meetings have included Food and Water Watch, Greenpeace, Physicians for Social Responsibility, and Code Pink. FFAN continues to press Senator Barbara Boxer's Environment and Public Works Committee for Congressional hearings on food safety due to Fukushima and also for the formation of an independent food-monitoring network so that all Americans may know the truth about radioactive fallout and how to protect our food supply.

Very special thanks to Mali Lightfoot, Melinda Woolf and Libbe Halevy for encouraging me to tell this story. Very special thanks also to Diane D'Arrigo and Cindy Folkers for their technical support.

16

CALL TO ACTION, GET INVOLVED

Supplemental Materials

Appendix A – Call to Action

Appendix B – Contact List

Appendix C – Fukushima Fallout Awareness Network (FFAN)

Appendix D - Media

Appendix E - Petition

Appendix F – Press Release

Appendix G – Glossary of Select Terms

Appendix A

Call to Action

So, what will YOU do to make things change? Be the change you wish to see!

Get into Action. Get Involved. Do your Piece.

Some suggestions of what you can do, today:

1. Check out the websites listed in the supplemental materials. (Appendix B)

2. Read more about FFAN. (Appendix C)

3. Join the FFAN Facebook group. (Appendix C)

4. Watch the videos on the media page. (Appendix D)

5. Sign the petition. (Appendix E)

6. Volunteer and offer your special gifts and services in support of FFAN or any of the organizations that are affecting change. You may find your special gift WHEN you volunteer!

7. Use your voice, loudly. Contact your local Congressperson or Senator; call the White House --(202) 456-1414.

8. Write letters to the editor to help educate others and ask the tough questions. www.dosomething.org

9. Continue to educate yourself, friends and family. Check out the media page for news sources for more information. (Appendix D)

10. Spread the word. Create a Facebook group. Share the messages of the current groups. But don't stop there. Create a video. Use the media, social media, your personal platform to help get the word out. Be creative.

11. Start an organization, group or movement in your area or find a way to support the ones listed in (Appendix B).

12. Be creative. The sky is the limit. If you can think it, you can do it. Don't second-guess yourself. Your opinion matters!

Appendix B

Contact List

My deepest gratitude to the following organizations for
their inspiration, assistance and expertise.
They have fought the good fight for many years and are all,
literally heroes in my book.

Beyond Nuclear

www.beyondnuclear.org

Citizens for Health

www.citizens.org

Ecological Options Network, EON

www.eon3.net

Fairewinds

www.fairewinds.com

Food and Water Watch

www.foodandwaterwatch.org

Greenpeace, USA

www.greenpeaceusa.org

Helen Caldicott Foundation, Nuclear Free Planet

www.nuclearfreeplanet.org

Nuclear Information and Resource Service, NIRS

www.nirs.org

Nuke Free

www.nukefree.org

Solartopia

www.solartopia.org

And please contact me for more information at
Fukushima Fallout Awareness Network

ffan@sonic.net

Appendix C

Fukushima Fallout Awareness Network (FFAN)

Mission

Fukushima Fallout Awareness Network is a coalition of groups and concerned citizens who share information and resources in response to the ongoing radioactive fallout of the Fukushima Daiichi nuclear disaster and then act to find solutions to the ongoing health threat of radioactive fallout. FFAN is committed to holding elected officials responsible for protecting the US food supply, as well networking with the people of Japan who are trying to protect theirs as well.

As of early 2012 core FFAN members include Nuclear Information and Resource Service, Beyond Nuclear, Citizens for Health, Ecological Options Network, Nuclear Free Planet, and concerned parents and grandparents.

Contact Kimberly Roberson for more information at:

ffan@sonic.net

Get Involved

Sign the Petition:

Fukushima Radioactive Fallout Food Safety Petition & Comments

http://www.change.org/petitions/urgent-fukushima-radioactive-fallout-food-safety-petition

Join the FFAN Facebook Group

https://www.facebook.com/groups/124537680994270/#!/groups/124537680994270/

'Silent Spring Meets The Long Hot Summer'

– Original Essay & Comments

http://nuclearfreeplanet.org/blogs/silent-spring-meets-the-long-hot-summer-kim-roberson.html

APPENDIX D

Media

(Please note that media links are active indefinitely. It may be necessary to do a separate search for them in your browser as they might become inactive.)

Videos and other media mentioned in the book.

Videos

• "Fukushima Solidarity in San Francisco"

There are two versions, one with and one without Japanese translation.

• "Fukushima Solidarity in San Francisco"

http://www.youtube.com/watch?v=dD-tDlQvyiw

• "Fukushima Solidarity in San Francisco" w/ Japanese translation

http://www.youtube.com/watch?v=WFSFDMiWVzU&feature=player_embedded

West Coast activists from FFAN deliver 7000 signatures from around the world to the San Francisco Japanese embassy in support of the Japanese mothers trying to prevent their government from shipping radioactive rubble from Fukushima to various locations in Japan, burning it, and dumping the ash in the ocean as a way of 'sharing the pain.' Created by Mary Beth Brangan and Jim Heddle of Ecological Options (EON) and FFAN.

Video

• "Lessons from Chernobyl"

http://www.youtube.com/watch?v=1hcBGSr9QGk

Dr. Dave DeSante is the founder of the Institute for Bird Population in Point Reyes, California. After the radioactive cloud from Chernobyl passed over the U.S. West

Coast in the spring of 1986 his research uncovered a severe die-off of young birds. Later, researchers Gould and Goldman duplicated his results with human mortality data from both the U.S. and Germany. In this in depth interview EON producers Mary Beth Brangan and Jim Heddle ask Dr. DeSante to explain his findings and their implications for today.

Video

• "Dial "M" for Meltdown" - by Brian Rich

http://fairewinds.com/content/dial-m-meltdown-brian-rich

Long time Fairewinds.com viewer and filmmaker Brian Rich has created a moving and high-energy chronology of nuclear power and its impact upon the world.

Nuclear HotSeat with Host and Producer, Libbe HaLevy

• Weekly Podcast

http://www.nuclearhotseat.com/

APPENDIX E

Fukushima Radioactive Fallout Food Safety Petition

By Kimberly Roberson

Why This Is Important

We are in the midst of an ongoing and seemingly incomprehensible radiation crisis at the Fukushima Daiichi nuclear power complex in Japan. Some are saying "this is not Chernobyl" and for that matter it may turn out to be worse. Greenpeace released a report on March 25 which ranks the radiation leaking from Fukushima to date to be at *Level 7* on the International Nuclear Event Scale (INES) which is the the highest level and the same as Chernobyl after that catastrophe. We must remember that this crisis is far from over and will have serious health effects for many innocent people, with young children and the elderly being particularly vulnerable. The National Academies of Science issued a 2006 report on radiation exposure that concluded that even *low levels of radiation* can cause human health problems, including cancer and heart disease. And 25 years after Chernobyl, the United Kingdom still maintains restrictions on sheep production in parts of Wales because radioactive cesium continues to contaminate grazing lands. Now we are learning that the water pumped in from the ocean to cool the Fukushima reactors is flowing back to sea mixed with deadly plutonium, endangering sea life too. Until workers can find a way to somehow stop, contain and store it, dangerous levels of radioactivity will continue to spread to the ocean and the biosphere. Particulates in the form of radioactive iodine and other radioisotopes from Fukushima have traveled across the United States as far as Massachusetts. *Increased radiation is now being detected in cows milk in Washington, Vermont, and other states.* While we continue to send our prayers and support to the people of Japan, it is clearly time to understand that this is a global crisis which will affect many nations, including the United States. Citizens in India, China, France and other countries are being told to carefully handle or not to eat large leafy green vegetables and some dairy products. Here in the US, closer downwind on the jet stream from Japan, we are STILL not receiving honest, accurate and consistent information from our government agencies. The FDA has announced that there is no need to test north Pacific fish for radioactivity, and the EPA announced scaling back monitoring of water and milk to quarterly testing only.

First, tell Congress and President Obama that we need to monitor all food and water imports from Japan, including the estimated annual 5 million gallons of bottled water, soft drinks and other non-alcoholic beverages containing water. Seafood shipments and other food products also are still being imported by the US.and they must be monitored immediately.

Next, tell Congress and President Obama that the Environmental Protection Agency must significantly expand the monitoring of air particulates, rain water, drinking water, and milk and to make the findings readily transparent and immediately available to the public.

Last, tell Congress and President Obama that the United States Department of Agriculture and the Food and Drug Administration should receive adequate funding for expanded food and water inspection both here and overseas and to communicate those findings immediately to the public. Congress must rethink our agricultural policies as well as international trade policies as they affect imports from other countries also trading with Japan.

SIGN THIS PETITION

http://www.change.org/petitions/urgent-fukushima-radioactive-fallout-food-safety-petition

APPENDIX F

Press Release

Fukushima Fallout Awareness Network (FFAN)
For Immediate Release,
Monday, November 7, 2011

Mothers Call for An Immediate Halt to the Burning of Radioactive Debris AND

Congressional Hearings Needed in the US to Monitor Milk and Food Safety:

"We Must Protect Our Children and Future Generations."

San Francisco and Washington, DC –

The Fukushima Fallout Awareness Network (FFAN) today delivered a petition to officials at the Japanese Consulate in San Francisco. The petition calls for an immediate halt to the burning of millions of tons of radioactive rubble and debris from the Fukushima Daiichi nuclear disaster on March 11. The signatures are from mothers and other concerned citizens in Japan and around the world who fear for their children's safety.

FFAN is extending their deepest sympathies to the Japanese people for the ongoing disaster of the March 11 earthquake and tsunamis and the Fukushima nuclear reactors. To avoid making the situation far worse, FFAN respectfully demands the immediate halt to the incineration of radioactive debris from the Fukushima Daiichi nuclear power disaster.

This just days after scientist Marco Kaltofen of Worchester Polytechnic Institute (WPI) presented his analysis of radioactive isotopic releases from the Fukushima accidents at the annual meeting of the American Public Health Association (APHA). Mr. Kaltofen's analysis confirms the detection of hot particles largely found in the western US but also contamination as far East as Boston, in addition to the extensive airborne and ground contamination in northern Japan due to the accident. FFAN, Mr. Kaltofen, and a large group of concerned experts and citizens in the US now believe that there is now not only a personal health issue in Japan, but a public health issue in the United States and Canada.

FFAN is also delivering letters to Senator Barbara Boxer and Dianne Feinstein today in San Francisco and Washington, DC which urgently request safety oversight and congressional hearings into the lack of milk, water, food and topsoil monitoring of cesium-137 (with a hazardous life of 300 years) and other dangerous radioisotopes now confirmed in California from the Japanese nuclear disaster. The Environmental Protection Agency has reduced their testing to four times per year. The University of California Berkeley School of Nuclear Engineering (UCBSNE) however, has been testing water, milk, topsoil and a variety of produce since the end of March and their findings are troubling to scientists and the general public alike.

"It is unfathomable to us that the Japanese government would intentionally allow this second wave of humanitarian and environmental crisis", said Kimberly Roberson of FFAN. "Burning the radioactive debris is not a solution, it needs to prevented from leaking into the environment and must be stored and carefully guarded well into the future." Mary Beth Brangan states "We have data from University of California Berkeley School of Nuclear Engineering showing elevated levels of cesium 137 in milk sampled from the San Francisco Bay area as recently as October 11th. These are radioactive particles that get into the body and can cause severe damage now and to future generations." Ms. Roberson insists, " We must have responsible monitoring and the findings need to be shared with the public as soon as possible to protect our children and their children as well."

Appendix G

Glossary of Select Terms

This is not intended to be a scientific reference, but is simply intended to help with clarification of some of the terms used in this writing.

Alpha particles

High-energy subatomic structure byproduct of nuclear power. Alpha particles can't travel very far and can be stopped by a piece of paper or skin. However, alpha can be dangerous once inhaled, ingested or through a cut in the skin.

"Atoms for Peace"

Program launched by a speech given by President Dwight D. Eisenhower post-World War II and part of a plan for turning atomic energy into energy production "too cheap to meter."

Beta particles

Electrons, a fraction of the size of alpha particles, are a byproduct of nuclear power. They travel farther and are more penetrating. They are a risk both to the internal and external parts of the body. Inhaling or ingesting a beta-emitting particle/radionuclide poses great risk.

Boron

Mineral used, among other things, to help reduce harmful radioisotopes created in nuclear power production. Used widely in the Ukraine during and after the Chernobyl nuclear meltdown, and at various nuclear waste facilities around the world to help reduce radiation exposure.

Chernobyl nuclear disaster

A nuclear power catastrophe which began on April 26, 1986, at the Chernobyl Nuclear Power Plant in the Ukraine. An explosion and fire released large quantities of radioactive contamination into the atmosphere which spread over much of Western USSR and Europe. Again, also rated as Level 7 on the International Nuclear Event Scale.

"China syndrome"

A catastrophic nuclear accident: a hypothetical accident in which the core of a nuclear reactor melts, allowing the radioactive fuel to burn through the floor of its container and straight down into the ground.

Fission

The spontaneous or induced splitting of an atomic nucleus into smaller parts, usually accompanied by a significant release of energy.

Fukushima Daiichi

A nuclear reactor complex in Japan which experienced catastrophic failure and meltdowns due to a chain of events following the great Eastern Japan earthquake and tsunamis of March 11, 2011. This event at Fukushima Daiichi is rated at a Level 7 on the International Nuclear Event Scale, the highest level possible on this scale, along with the Chernobyl disaster.

Gamma rays

The most penetrating type of radiation that can only be stopped by thick lead blocking their path. Radioactive cesium and iodine are gamma emitters that are byproducts from nuclear power.

Greenpeace

A non-governmental (NGO) organization with offices in over 40 countries, with its headquarters in Amsterdam. Greenpeace states its goal is to "ensure the ability of the Earth to nurture life in all its diversity."

Half-life

The time required for half the nuclei in a sample of a specific radioactive isotope to undergo radioactive decay.

Hot particles

Microscopic pieces of radioactive material which are dangerous due to their ability to become lodged in a person's body and thereby deliver a concentrated dose of radiation to a small area over a long period.

Ionizing radiation

High-energy radiation capable of producing ionization in substances through which it passes; the type of radiation produced by nuclear power.

Low Level Radioactive Waste

A broad category, which, in the U.S., includes booties and gloves from nuclear medicine as well as all commercial nuclear waste except irradiated fuel from nuclear reactors and more.

Meltdown (see "China Syndrome")

NGO

Non-governmental organization

Nuclear power

The use of sustained nuclear fission to generate heat and electricity.

Radiation

A form of energy that passes through space or tissue in the form of waves.

Radioactive isotope

An isotope having an unstable nucleus that decomposes spontaneously by emission of a nuclear electron or helium nucleus and radiation, whereby achieving a "stable" nuclear composition.

Renewable energy

Forms of energy including; wind, solar, geothermal and select biomass that are alternatives to traditional polluting energy methods such as gas, coal and nuclear.

Spent fuel

Fuel in nuclear power production no longer considered usable. It remains extremely dangerous for millions of years.

Trans-generational DNA damage

to DNA passed on to future generations.

Uranium

A radioactive heavy metal used in nuclear power production.

Uranium mill tailings

Waste produced by the extraction and mining of uranium. A deadly byproduct of nuclear fuel production that remains hazardous for millions of years.

Ward Valley radioactive dump site

Proposed "low-level" nuclear dump in Ward Valley, California, in the 1990s. Activists including Native American Tribes fought for over 10 years to defeat it.

AFTERWORD

Since completion of this book there has been a very significant development. The San Onofre nuclear reactor in southern California has been closed for several months due to safety issues. Unit 2 was taken out of service for a planned outage on Jan. 9, 2011. Unit 3 has been shut down since Jan. 31. Weakened fuel rods were discovered (rods which had recently been installed at great expense as a system upgrade). Senators Barbara Boxer and Dianne Feinstein have vowed to keep San Onofre shut down until all of their safety concerns have been properly addressed. As you will remember from earlier in the book, this nuclear reactor sits directly on the beach of the Pacific Ocean, on and near active earthquake fault lines — a situation very similar to Fukushima Daiichi. Dedicated activists who have been working for years to close San Onofre due to earthquake and structural concerns have been motivated more than ever to keep it closed once and for all. Although the work really never ends <u>due to the fact that the radioactive waste from the reactor will remain hazardous for tens of million years, however closing the operation of the reactor is a good start.</u> Once again, remember that California produces over 450 varieties of produce and is the largest producer of dairy in the U.S. You can do the math and imagine the consequences of a nuclear meltdown at San Onofre.

The nation's top nuclear regulator, Nuclear Regulatory Chairman commissioner Gregory Jazcko, recently visited the site along with California lawmakers and has promised to keep the reactors closed offline until all safety issues are resolved. I had the timely good fortune to speak with one of Senator Barbara Boxer's top aides recently upon his return from a visit to San Onofre with Chairman Jazcko and was very heartened to hear again that Senator Boxer is determined to protect her state. We at Fukushima Fallout Awareness Network are asking for one further step: to keep San Onofre closed permanently, even if it means rolling energy blackouts and brownouts during the upcoming summer months. A blackout sure the heck beats the alternative of another Three Mile Island, Chernobyl, or Fukushima Daiichi catastrophe.

What happens in California doesn't usually stay here because the state has a reputation for initiating environmental reforms that spread to the rest of the nation and even the world. People look to the Golden State for leadership on environment and health safety policy. The San Onofre issue offers a very unique opportunity to begin a more comprehensive transition to wind and solar power now.

Meanwhile, back in Japan the problem is far from over. Japanese officials are continuing with their plan to incinerate tens of millions of tons of earthquake and tsunami rubble. Much of it contains harmful chemicals and radiation from Fukushima Daiichi. And as if that was not bad enough, the situation at Daiichi's reactors number 3 and 4 are still critical and another earthquake could release harmful plutonium into the jet stream. Why is there still so much silence around the world during this international crisis at Fukushima Daiichi? When will the status quo change? When will it no longer be acceptable to attempt to utilize a form of energy over which we have no control? So many questions …The answers often come from concerned citizens getting involved. Dear reader, this is your call to action. Stay safe and be well.

ACKNOWLEDGEMENTS

It really is amazing to think of how much goes into a project such as this, and how the involvement of many people is integral to making it "fly".

I want to thank in particular one woman who gave so much of her talent, heart and soul to help bring this story to you. For nearly one year it was an upstream struggle to raise awareness to the Fukushima radioactive fallout food safety issue, however Melinda Woolf's resolve was total and her message clear: the pain and suffering of Fukushima Daiichi would not be in vain, that a wide audience should be able to read and know the story of the need to protect our food supply and future generations. Melinda has truly been like an angel — one who helped to put my experience into a spiritual context and therefore really completed the journey thus far….

Mali Lightfoot, for asking for me to write the story of the food safety petition, which was posted on her and Dr. Helen Caldicott's Nuclear Free Planet website. It planted the seed for this book. Without her, you wouldn't be reading this! Thank you Mali!

And special thanks to Libbe Halevy for reading the story online and for encouraging me to turn it into an ebook. The thought had not dawned on me. She is another angel who arrived somewhat unexpectedly and is so appreciated.

Cindy Folkers and Diane D'Arrigo have worked tirelessly and for many years protecting the public from the dangers of radioactive fallout. I remember Diane from the 1990's when we went together to a meeting at the White House's Old Executive Office Building to demand an end to the proposed Ward Valley dumpsite in California. Cindy and I only met this past year but it didn't take long to understand her compassion and intelligence. All of us owe them a debt of thanks for their commitment to nuclear awareness and public safety. They not only helped with the technical aspects of writing and fact checking this book, but offered hours of their time, at all hours of the day. They are true heroes in my book, pun intended!

Gail Payne is a nuclear activist in her own right who knew instinctively what it was I wanted in a book cover for this piece. I want to thank her for her extremely committed and generous support of not only putting a face to this story, but also the Fukushima Fallout Awareness Network (FFAN) by designing our logo and website.

Very special thanks to the members of FFAN who were not already acknowledged above: Rachel Gertrude Johnson, Mary Beth Brangan, Jim Heddle, Jim Turner for their lifelong commitment to protecting the public and to all future FFAN members as well!

And last but certainly not least, I want to thank my husband and life partner Tim for his compassion and support through a very long and sometimes difficult process. He often reads the compass better than I! And for understanding how life takes on unexpected turns and our need to adapt and rise to the occasion. Much love to you!

www.ingramcontent.com/pod-product-compliance
Lightning Source LLC
Chambersburg PA
CBHW060644210326
41520CB00010B/1731